BEST 10

21st ASIA-PACIFIC INTERIOR DESIGN AWARDS

第二十一届亚太区室内设计大奖入围及获奖作品集

深圳市艺力文化发展有限公司 编

· 广州 ·

图书在版编目（CIP）数据

第二十一届亚太区室内设计大奖入围及获奖作品集：英汉对照 / 深圳市艺力文化发展有限公司编 . — 广州：华南理工大学出版社，2015.2
　ISBN 978-7-5623-4476-6

Ⅰ . ①第… Ⅱ . ①深… Ⅲ . ①室内装饰设计 - 亚太地区 - 图集 Ⅳ .
① TU238-64
中国版本图书馆 CIP 数据核字（2014）第 274897 号

第二十一届亚太区室内设计大奖入围及获奖作品集
深圳市艺力文化发展有限公司　编

出 版 人：韩中伟
出版发行：华南理工大学出版社
　　　　　（广州五山华南理工大学 17 号楼，邮编 510640）
　　　　　http://www.scutpress.com.cn　E-mail: scutc13@scut.edu.cn
　　　　　营销部电话：020-87113487　87111048（传真）
策划编辑：赖淑华
责任编辑：王　岩
印 刷 者：深圳市福圣印刷有限公司
开　　本：787 mm×1092 mm　1/8　印张：30.75
成品尺寸：290 mm×350 mm
版　　次：2015 年 2 月第 1 版　2015 年 2 月第 1 次印刷
定　　价：798.00 元

版权所有　盗版必究　　印装差错　负责调换

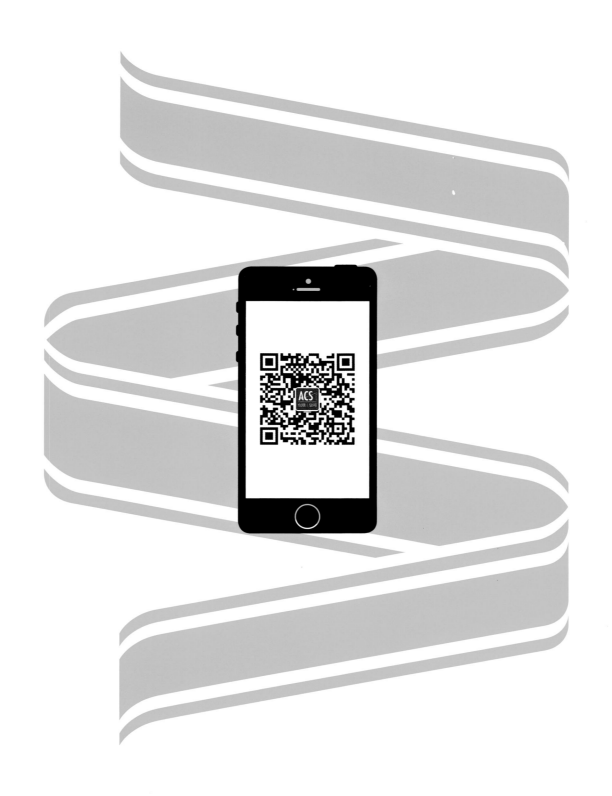

扫描二维码即可开启
前所未有的动态图书阅读体验!

电脑版 *www.acs.cn* 与 手机版 *m.acs.cn* 同步上线

APIDA

PREFACE
序 言

Antony Chan
Chairman
Hong Kong Interior Design Association

Over the past 20 years, the Asia Pacific Interior Design Award witnesses the design evolvement in the Asia Pacific Region that has now achieved an international recognition. Entries from different countries demonstrate significant and encouraging design progress. Surely, APIDA has become a very important platform to honor outstanding design achievement and thereby encourage excellence in all facets of interior design.

APIDA 2013 has once again achieved new standards and will no doubt set design trends in the region. I am convinced, in the future, with continuously support of all participating designers, judges, and also sponsors, co-organizers and volunteers, APIDA shall rise to new heights.

Wesley Liu
Chairman
APIDA 2013

In celebration of our continued success for the 21st Asia Pacific Interior Design Awards, on behalf of HKIDA, I would like to take this opportunity to express our deepest gratitude to all of our executive committee, members and supporting organizations. Reaching this milestone has only been possible through the continued support from all of you.

It is my great pleasure to chair the APIDA as well as the East Gathering event this year. Together, we have witnessed year on year, an increasing number of participants and an ever-growing pool of young, upcoming talents within the Interior Design industry all across the Asia Pacific. With our concerted efforts in raising professional standards and the exchange program with Asia Pacific countries, we have been successful in bridging the design cultures as well as promoting each of the winners work overseas. Ultimately, it is our wish that APIDA will not only be the most recognized award, but also a powerful platform in which to promote Asia Pacific Designers.

With the APIDA's reputation as one of the most important Interior Design awards in the region, we have been delighted to have received nearly 600 entries from the Asia Pacific area. There has also been a drastic increase in the number of overseas submissions this year, which is an encouraging and remarkable result of our tri-city partnership with Japan and Korea, achieved through our East Gathering as well as other regional networking efforts.

To ensure the high quality and recognition of the APIDA awards, we have worked tirelessly in search of highly respected Designers across different countries and various fields within the Interior Design industry to join our panel of judges. Taking things a step further, instead of the previous one-day intensive judging process, this year saw the judging schedule split into 2 half-days, which allowed our judges to scrutinize over the shortlisted projects. This not only granted extended viewing time but also established a fairer assessment for each entry.

I would like to take this opportunity to express our heartfelt thanks to all our overseas and local judges for their valuable time, genuine insight and supportive aspirations, as well as all the young talents whom had shared with us their successful projects through their submissions. The passion and dedication showcased in your works serves as a ray of hope towards the industry, foretelling a brighter and greater future for Asia Pacific Interior Design.

On reaching our 20th anniversary, I would like to take this wonderful opportunity to express our utmost gratitude to all of our current and past sponsors for their tremendous support towards the APIDA. Your generous and combined support has helped build the foundations towards the success and growth of our industry. We thank you all from the bottom of our hearts and look forward to your continued support in the coming years!

Last but not least, we wish to extend our congratulations to all the winners this year. These awards recognize the excellence and contributions you have made towards our industry. It is our sincere hope that this helps motivate and inspire other Designers to develop and push their creativity further, creating works that will serve to inspire others in years to come.

JUDGES 评委

Federico Delrosso
Architect of Federico Delrosso Architects

"Architecture must have a soul — it's own soul that is echoed in every detail. Light and shadow must be delicately balanced to reach the unique and magic point. It should have an independent voice from the people who made it and those that live in it — created by man and for man. This is my mission…" F.D.

I believe that true architecture cannot stop at the surfaces of a home, but should also develop without a break, like a Moebius strip, from the outside to the interior. In my project the exterior always finds its 'natural' homolog within. But more than a simple correspondence (the void that reflects the solid) I enjoy working on two fundamental links: light and natural materials." F. D.

An architect with an international range (France, Switzerland, USA, Montecarlo, Turkey), but with a solid Italian grounding, Federico Delrosso (born in Biella in 1964; graduated from the Milan Polytechnic in 1996) within a few years has gained wide experience in the design of a very wide range of types of architecture: private housing, restaurants (for the catering industry), a rehabilitation centre, office buildings... numerous projects of interior design but also industrial design.

His early experience in the field has consolidated a talent and a distinctive sensibility for spaces, perceived in his projects as an intangible quality which Federico Delrosso calls "the emotional fourth dimension". In other words, in his interiors as well as his architectures, the result of the masterly relationship between three-dimensional spaces and the elaboration of surfaces is always a sense of pleasant balance, typical of fine architecture.

Joanne Cys LFDIA
Dean: Teaching and Learning. Division of Education, Arts and Social Sciences
Associate Professor in Interior Architecture of South Australia

Joanne Cys LFDIA is an Executive Board Member (2011-2013) of the International Federation of Interior Architects/Designers (IFI) and she is Australia's representative to the Global Design Network (GDN) and the Asia Pacific Space Designers Association (APSDA). At the University of South Australia she is Associate Professor in Interior Architecture and Dean of Teaching and Learning.

Joanne was a founder of the Australian Interior Design Awards program and has been its jury convenor since its inception in 2004. She was an invited member of IFI's International Scientific Committee for the international 2011 Design Frontiers: Interiors Entity (DFIE) Global Symposium and is Co-Chair of the Global Interior Educators Open Forum (GIEOF).

Joanne has been invited to speak at international and national conferences, curate design exhibitions and is regularly invited to write for professional design journals. She has published over 50 academic papers in scholarly journals, conference proceedings and chapters in edited books.

Makoto Tanijiri
Architect of Suppose Design Office

Makoto Tanijiri was born in 1974. In 2000, he started Suppose Design Office, an architectural design firm in Hiroshima. His work covers a broad range of areas including designing houses, business spaces, site frameworks, landscapes, products, and art installations. He has received many professional recognitions such as the JCD Rookie Award, and his design for Heiwa-ohashi Pedestrian Bridge Design Proposal Competition was chosen as a finalist. He is also a part-time professor at Anabuki Design College.

Jooyun Kim
Professor at Hongik University & Design Director of Studio Button

Jooyun Kim is a professor at Hongik University in Seoul, Korea. In 1984, He got his bachelor's in Architecture at Hongik University. In 1991, Jooyun finished his master's in Interior Design at Cornell University and got a PhD. in Architecture at Kookmin University in 2002. In 2004, He worked as a research professor at Pratt Institute in the U.S. Jooyun had served as the president of KOSID (Korea Society of Interior Architects/Designers) 2011-2012 as well as the executive board member of IFI (International Federation of Interior Architects/Designers) 2007-2011. He also works as the design director at a Seoul-based design firm Studio Button.

Jooyun gave numerous lectures worldwide. He spoke at 1998, 2000 and 2005 APSDA (Asia Pacific Space Designer's Association); 2006 IFI roundtable; 2007 and 2010 LuXun Academy of Art in China; 2008 IIID Conference in Ahmadabad; 2008 U.Design in Monterrey, Mexico; 2010 Decorex-10 in Johannesburg; 2010 Desain.ID in Jakarta and 2011 Maison& Object in Paris. He also served as a judge internationally at the UNESCO Creative City in 2010; Singapore Design Awards in 2009, 2010 and 2011; Plascon Prism Award in 2010; IIDA Award in 2009 and Seoul Design Olympiad Award in 2009.

Notably, he was selected as the designer of the year in 2006 by the Ministry of Construction & Transportation and received the Galmae Award from KOSID in 2008. The Galmae Award is given to those who contributed the most to the development of Korean interior design/architecture in the last 10 years. Jooyun published Jooyun Kim (2010), Augmented Space through Media (2010), and Invited Architecture Masters (2009).

Yul Lin Wang
President of Chinese Society of Interior Designers (CSID)

· Current President of CSID (Chinese Society of Interior Design)

· Current vice president of TDA (Taiwan Design Alliance)

· Current MPI Design owner

· Selected by Taipei city in 2012 as examine committee of the 2016 World Design Capital.

· Current assistant professor of interior design department at Chung Yuan Christian University.

· Current supervisory committee of tourism hotel quality of Taiwan Tourism Bureau.

· Current accreditation committee of tourism hotel ranking of Taiwan Tourism Bureau.

· Member of IFI (International Federation of Interior Architects/ Designers)

Winning Records

· 2013 ChanYan- Selected for German Design Award 2014

· 2012 ChanYan- Red Dot Architecture category- Red Dot Special Design Award.

· 2012 Pureness and Peacefulness- Red Dot Architecture category- Red Dot Special Design Award.

· 2010 Taiwan Top 10 Designers.

· 2010 YueTai- TID (Taiwan Interior Design)Award- Temporary Architecture TID Award.

David Tsui
Principal of HASSELL

David Tsui is a Principal at HASSELL, an international design practice with 14 studios in the world. David has more than 20 years of experience in interior design, and has led the design for many important projects across different sectors and industries in the region, including hospitality, residential and commercial space. These projects can be found in Indonesia, South Korea, Hong Kong, and many cities in China.

David is recognised for his creativity and sensitivity to the client's need. He has received numerous awards and recognition in Hong Kong and China. He was the winner of the Australia and Hong Kong Interior Designer Association (Hospitality Interior Design) Award. He also won first prize at the Interior Design China Hospitality Design Awards, Outstanding Greater China Design Award and was a Top Ten Outstanding Designer in Hong Kong in the year 2010 and a Top 10 Asia Pacific outstanding hotel designers in the year 2012.

Greg Farrell
Executive Director of Aedas Interiors Limited

Greg Farrell brings over 26 years of experience with specialty in hospitality design leadership and project management to Aedas Interiors. Originally from New Zealand, Greg has practiced in various international locations, starting his hotel design career at Richmond Design in London and having worked on projects from Australia to SE Asia, China, Middle East and Europe. As Executive Director, he leads the Aedas Interiors global hospitality practice.

His focus in early-phase hotel planning and keen interest in guestroom planning has proven to resonate with both hotel operators and owners. He aspires to create interactive and intuitive spatial experiences. His dynamic guestroom plans result in spaces that add interest with a sense of discovery whilst incorporating flexible circulations and divisions that can allow for optimal privacy or the option to open up the entire room.

Greg's breadth of knowledge gained from numerous 3-5 star hotel, luxury resort projects include hotel brands such as, Jumeirah Resorts, Langham, Langham Place, Radisson Blu, Hilton DoubleTree, Hotel Indigo by IHG and Ritz Carlton to name a few.

Ed Ng
Co-founder & Director of AB Concept Ltd.

Ed Ng's designs reveal his genuine respect for history, culture and arts. While his work draws inspiration from the local culture, he reinterpretstraditional motifs or cultural touchstones to add a contemporary twist. His designs, richly detailed with a couturier-like precision, always contain an element of surprise – from an unexpected use of colour to a bold use of fabric– to ensure each project evokes a sense of bespoke beauty.

Innovative and creative, Ed collaborates with clients to ensure the final product both fulfils his conceptual vision and serves their functional needs.

Ivan Dai Nap Kwan
Partner of LRF Designers Ltd.

Qualifying in 1977 at the renowned Swire School of Design at Hong Kong Polytechnic University. He began his career working within a multidisciplined architectural practice where he was primarily involved with interior design projects. He became a member, then fellow in 1994, of the Chartered Society of Designers. He continues to play an active role in design education and at a broader professional level – having been elected Chairman of the Interior Design Association (Hong Kong) from 1998 to 2000 and Founder Member of the Hong Kong Design Centre. Dai's solid experience helming five-star hotel work through sound project management, contract administration and financial control is largely responsible for LRF Designers Ltd.'s successful track record. Furthermore, he has been appointed by the HKCAAVQ as a specialist in 2001, 2008 and 2011, and Vice Chairman of the Asia-Pacific Hotel Design Association in 2010.

Prof. Cees de Bont
Dean & Swire Chair Professor of Design
School of Design
The Hong Kong Polytechnic University

Prof. Cees de Bont is the Dean of the School of Design, the Hong Kong Polytechnic University. He obtained a MA in Consumer Psychology from Tilburg University and a PhD in Industrial Design from Delft. Prof. Cees de Bont has a good mix of academic and industrial experience. From 1997 to 2005, he was head of Marketing Research and Strategy at Philips Domestic Appliances and Personal Care. In this function, he was responsible for generating and utilizing market information for the formulation of brand strategies and marketing plans for the company. From 2005 onwards, he serves as a dean in Delft University of Technology and after that Hong Kong Polytechnic University. Cees chaired Dutch Innovation Centre for Electric Road Transport which is a platform for research on electric mobility D-Incert.

Kinney Chan
Founder of Kinney Chan & Associates

Kinney obtained his BA in Interior Design from Demonfort University, UK. Prior to returning to Hong Kong and founding KCA, he spent few years working in Manchester.

Kinney is the former chairman of the Hong Kong Interior Design Association and the former director of the Hong Kong Design Centre. He strives to increase public awareness of interior design as well as the development of the creative industry in Hong Kong.

His projects have won numbers of international awards, including Best Of the Year Honoree 2012; Hong Kong Ten Outstanding Designers Award in 2012; iF Communication Design Award in 2009; Asia Pacific Interior Design Awards in 2001,and for many times awarded as one of the Best Interior Designer Worldwide in Andrew Martin International Awards.

CONTENTS 目录

GOLD MEDAL 金奖

- 002 / THE COMMUNE SOCIAL
  食社
- 006 / XI'AN WESTIN MUSEUM HOTEL
  西安威斯汀博物馆酒店
- 011 / NANCHANG INSUN INTERNATIONAL CINEMA
  南昌新华银兴国际影城
- 014 / D.R.HOME
  设计共和·家
- 018 / MADAM CHANEL
  香奈儿女士
- 022 / CITY CROSSING
  华润中心
- 026 / GEHUA YOUTH AND CULTURAL CENTER
  歌华营地体验中心
- 030 / ADIDAS GREATER CHINA HEADQUARTERS
  阿迪达斯大中华区总部
- 034 / UMIX TOP UNDERWEAR MULTI-BRAND STORE
  Umix 高端内衣综合品牌店

FOOD SPACE 用餐空间

- 040 / FARINE
  Farine 面包房
- 044 / W HOTEL FEI ULTRA LOUNGE
  W 酒店休息室
- 050 / WADAKURA IN PALACE HOTEL
  皇宫酒店 Wadakura 餐厅
- 054 / JIN & INA
  牛郎 & 织女
- 059 / HAZEL & HERSHEY CAFE
  Hazel & Hershey 咖啡
- 062 / ENJOY COFFEE
  享入啡啡
- 068 / SAL CURIOSO
  Sal Curioso 餐厅
- 072 / CAPO RESTAURANT
  Capo 餐厅
- 076 / KAMPACHI AT THE TROIKA
  Kampachi 餐厅

HOTEL SPACE 酒店空间

- 084 / REGENT PHUKET CAPE PANWA
  普吉岛攀瓦角丽晶酒店
- 090 / INTERCONTINENTAL HONG KONG (BALLROOM, LOBBY AND LOBBY LOUNGE)
  香港洲际酒店（宴会厅、大堂和大堂酒廊）
- 094 / MANDARIN ORIENTAL PUDONG SHANGHAI
  上海浦东文华东方酒店
- 100 / HAITANG BAY NO.9 RESORT SANYA
  三亚海棠湾 9 号度假酒店
- 108 / HOLIDAY INN RESORT, CHANGBAI MOUNTAIN
  长白山万达假日度假酒店
- 112 / AQUA SALON
  时尚精品饭店会所酒店
- 116 / HUBIN SPRING SEASON HOTEL
  湖滨四季春酒店
- 130 / WUHAN LIGHT TEXTILE EMPLOYEE SANATORIUM 4TH PHASE - THOUSANDS STREAM TO THE MANSION
  武汉市轻纺职工疗养院 4 期·千水归堂
- 134 / ALOFT KL SENTRAL
  中央车站雅乐轩酒店

LEISURE & ENTERTAINMENT SPACE 休闲/娱乐空间

- 142 / TIMES BUND CLUB HOUSE
  时代会所
- 148 / DFS PSC LOUNGE
  DFS 铂金服务俱乐部会所
- 154 / TIANJIN INSUN LOTTE CINEMA
  天津阳光乐园电影院
- 158 / REME CLUB
  REME 会所
- 164 / MOUNT DAVIS YOUTH HOSTEL
  戴维斯山青年旅馆
- 169 / THE GYM BOX
  健身拳击俱乐部
- 172 / CLUB AXIS THE WINGS
  Axis 会所
- 178 / AKOZO SALON & SPA
  中国南京 AKOZO 资生堂沙龙会所
- 184 / ENZO CLUB
  ENZO CLUB

LIVING SPACE 居住空间

- 192 / RESIDENCE C
  住宅 C
- 196 / FLOWING INK
  墨方
- 200 / LUKE HOME
  卢克的家
- 204 / SKETCH PERSONALITY
  速写延伸个性
- 208 / READING SEQUENCE
  阅读空间的顺序
- 212 / THIRD DERIVATIVE
  衍伸
- 216 / HOUSE IN SHATIN
  沙田小屋
- 220 / JETTY HOUSE
  支架屋
- 224 / 香港仔鸭脷洲"船屋"公寓

SAMPLE SPACE
样板房空间

- 231 / 海宁样板房
- 236 / 绿地成都东村 468 公馆样板房
- 240 / Yinyi Ishakes 酒店式公寓
- 244 / 隐山居
- 248 / 京投银泰宁波东钱湖悦府一期高端别墅——枫丹白露悦府会
- 254 / 万科松山湖中心别墅
- 259 / 万科悦湾 A2 复式洋房
- 264 / 雍雅山 15 号住宅
- 270 / 宁波华润置地集团国际社区复式样板房

INSTALLATION & EXHIBITION SPACE
设施 / 展览空间

- 276 / TIMES · BLOOM
  时代 · 花生
- 280 / SWIRE PROPERTIES LOUNGE AT ART BASEL IN HONG KONG
  太古地产贵宾厅——香港巴塞尔艺术展
- 284 / CHONGQING FLOWER & CITY SALES OFFICE
  重庆花城售楼处
- 288 / IDEA DOOR
  多维门
- 292 / EYE TO EYE: JOCKEY CLUB SOCIAL DOCUMENTATION ROVING EXHIBITION
  另眼·相看：赛马会社会纪实主题巡回展
- 296 / NORTH GARDEN SALES PAVILION & GALLERY
  北京天洋北花园销售中心
- 300 / DIALOGUE — ZHONGJUN SHANGCHENG SALES CENTER
  《对话》——中骏商城销售中心
- 304 / HILLS SALES CENTRE
  西山庭院售楼处
- 308 / YUE HOUSE
  京投银泰宁波东钱湖悦府一期高端私人会所售楼处——悦府会

PUBLIC SPACE
公共空间

- 316 / CHAPEL
  Chapel 教堂
- 320 / DALIAN INTERNATIONAL CONFERENCE CENTER
  大连国际会议中心
- 324 / CHONGQING VANKE YUEWAN SALES CENTER
  重庆万科悦湾销售中心
- 328 / DESIGN REPUBLIC DESIGN COMMUNE
  设计共和设计公社
- 332 / SPRING
  春天
- 336 / G CLINIC
  G 诊所
- 338 / SPATIAL REORGANIZATION OF THE UNIVERSITY LIBRARY IN CUHK
  香港中文大学图书馆空间改造
- 342 / GREENLAND M SALES CENTER
  绿地·M 中心售楼处
- 346 / ANHUI · ANQING FUCHUN ORIENT SALES CENTER
  安徽·安庆富春东方销售中心

WORK SPACE
办公空间

- 354 / XIAMEN HIMALAYA DESIGN CO., LTD. OFFICE
  厦门喜玛拉雅设计装修有限公司办公室
- 358 / THE MOMENT OF CHANGE
  改变的时刻
- 362 / MASAN SINGAPORE OFFICE
  马山集团新加坡办公室
- 366 / SUMITOMO MITSUI BANKING CORPORATION (SMBC) SHUKUGAWA BRANCH
  三井住友银行夙川分行
- 372 / SPEEDMARK
  立通
- 376 / NANSHAN ZHONGTAI TIANCHENG OFFICE
  南山中泰天成办公室
- 380 / GCL SHANGHAI HEADQUARTER
  GCL 上海总部
- 384 / XIAMEN FENGYA DESIGN CO., LTD. OFFICE
  厦门风亚建筑设计顾问有限公司办公室
- 388 / HUAKUN INVESTMENT
  华坤投资
- 392 / MERCK SHARP & DOHME TAIPEI OFFICE
  默克沙东台北办公室
- 396 / SAMLEE OFFICE
  仕其商贸有限公司（台湾、香港）广州办公总部

SHOPPING SPACE
购物空间

- 402 / INTERIOR DESIGN FOR TEA TAO (BROAD AND NARROW ALLEY STORE)
  茶道室内设计
- 407 / ESTNATION FUKUOKA
  ESTNATION 福冈店
- 410 / UM COLLEZIONI FEMALE TOP FASHION MULTI-BRAND STORE
  UM Collezioni 高端女装综合时装店
- 414 / BODHI VENUE
  柏地广场
- 420 / LOUNGE BY FRANCFRANC
  Francfranc 休息室
- 426 / APPLE & PIE
  苹果 & 馅饼
- 430 / GUANGZHOU TEXTILE EXPO CENTER STORE
  广州纺织博览中心商铺
- 436 / MERCATO FRESCO
  梅尔卡托弗雷斯科
- 439 / DFS LE SALON
  DFS 沙龙
- 442 / LANGFANG DEFA CLASSICAL FURNITURE EXPERIENCE PAVILION
  廊坊德发古典家具体验馆

STUDENTS' WORKS
学生作品

- 450 / A FOLDING ARCHITECTURE: THE CENTER OF HOSPICE AND PALLIATIVE CARE
  折叠式建筑：临终关怀和姑息治疗服务中心
- 454 / FLOATING COMMUNITY - H2O HOTEL
  漂浮社区——H2O 酒店
- 456 / STUDARY
  墨坊
- 458 / MOVABLE BEACH
  可移动的海滩
- 460 / BAND SOUND CENTER
  乐队音乐中心
- 464 / CHURCH
  教堂
- 466 / YOUTH HOSTEL
  青年旅社
- 470 / PLEASURE LEARNING
  寓学于乐
- 472 / REFRESHING EXPERIENCE IN MIND & BODY HEALTH CLUB
  身心健康的全新体验俱乐部

21st APIDA
GOLD MEDAL
金 奖

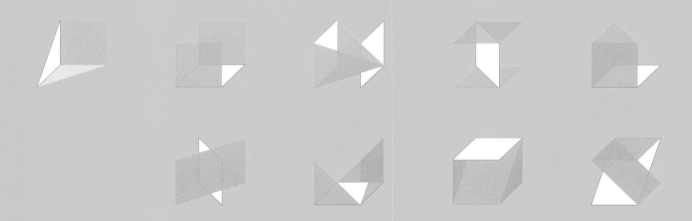

● FOOD SPACE　　GOLD MEDAL
用餐空间　　　　金奖

THE COMMUNE SOCIAL

食社

Design Agency / Neri&Hu Design and Research Office

The Commune Social located at the Design Republic Design Commune is a new tapas restaurant designed by Neri&Hu Design and Research Office of Shanghai. The food concept is a fresh and modern take on Spanish tapas, where the small dishes served are refined interpretations of the tapas genre with an international twist.

The design concept of the space is to emphasize the "sharing" idea of eating tapas, and the four distinct areas in the restaurant (tapas bar, dining room, dessert bar, and the secret bar) are meant to provide different flavors and ingredients for diners to "share", like they do the food. The diners are encouraged to move from having drinks at the Secret Bar to the tapas bar or the dining room for dinner, then end up at the Dessert Bar to finish the meal. Each room serves something different, and each room evokes a different atmosphere. ●

Design Team / Lyndon Neri, Rossana Hu and others
Location / China
Area / 495 m²
Client / Jason Atherton, Loh Lik Peng

社会公社位于设计共和公社内,是由上海如恩设计研究室设计的一间餐厅。餐厅内的食品理念承袭自西班牙餐前小吃的新鲜和时尚,小碟子盛放的精致美食诠释了国际风

格的餐前小吃类型。

空间的设计理念是为了强调"分享",餐厅中有4个明显的分区(餐前小吃吧、用餐室、餐后甜点吧和秘密吧),提供各种风味的食物给就餐者"分享"。餐厅鼓励就餐者从喝饮品的秘密吧转移到餐前小吃吧或者是就餐室,最后在甜点吧结束就餐。每个空间都供应一些不同的东西,每个空间都有不同的氛围。

- Hotel Space 酒店空间
- GOLD MEDAL 金奖

Design Team / Lyndon Neri, Rossana Hu and others
Location / China
Area / 100,000 m²
Client / YunGao Hotel (Group)

XI'AN WESTIN MUSEUM HOTEL

西安威斯汀博物馆酒店

Design Agency / Neri&Hu Design and Research Office

In an ancient capital of China, Neri&Hu Design Research Office's design of the Westin in Xi'an emerges as a tribute to both the city's importance as a hub of burgeoning growth in the region, as well as its long standing status as a cradle of Chinese civilization. With 3,100 years of history embedded in the layers of the city, Xi'an is not merely a formidable backdrop to the building itself but has provided the architects with design inspirations that inextricably link its past to its present and future.

Arriving in Xi'an, one is immediately struck by the fortress-like expanse of its enveloping city walls, and the architecture of the Westin takes cues from this heavy monumentality. In view of its urban context, the dark stucco and stone clad building blocks adopt the profile of vernacular Chinese architecture.

　　如恩设计研究室所设计的西安威斯汀博物馆酒店，旨在向西安这个重要的区域经济中心城市致敬，同时体现其在中国文明史上所占有的重要地位。3100多年的历史融入城市的脉络中，西安不单是建筑所在地，其过去、现在与未来有着千丝万缕、密不可分的联系，同时也为建筑师带来设计灵感。

　　到达西安的人们会立刻被其壮观的城墙所震撼。威斯汀博物馆酒店的整体设计从这种庄严的厚重感中吸取灵感，所用材料选择了深色的灰泥和石块，完美融合了本土文化与现代元素。

MUSEUM HOTEL

● LEISURE & ENTERTAINMENT SPACE　　GOLD MEDAL
休闲/娱乐空间　　金奖

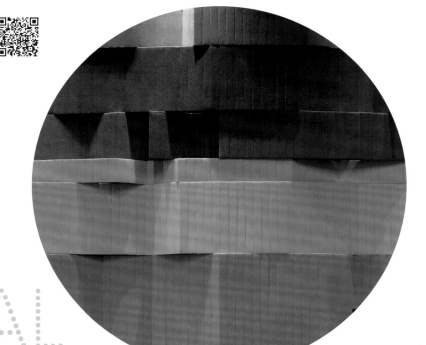

NANCHANG INSUN INTERNATIONAL CINEMA

南昌新华银兴国际影城

Design Agency / One Plus Partnership Ltd.

Design Team / Virginia Lung, Ajax Law
Location / China
Area / 7,285 m²
Client / Hubei Insun Cinema Film Co., Ltd.

INTERNATIONAL CINEMA　011

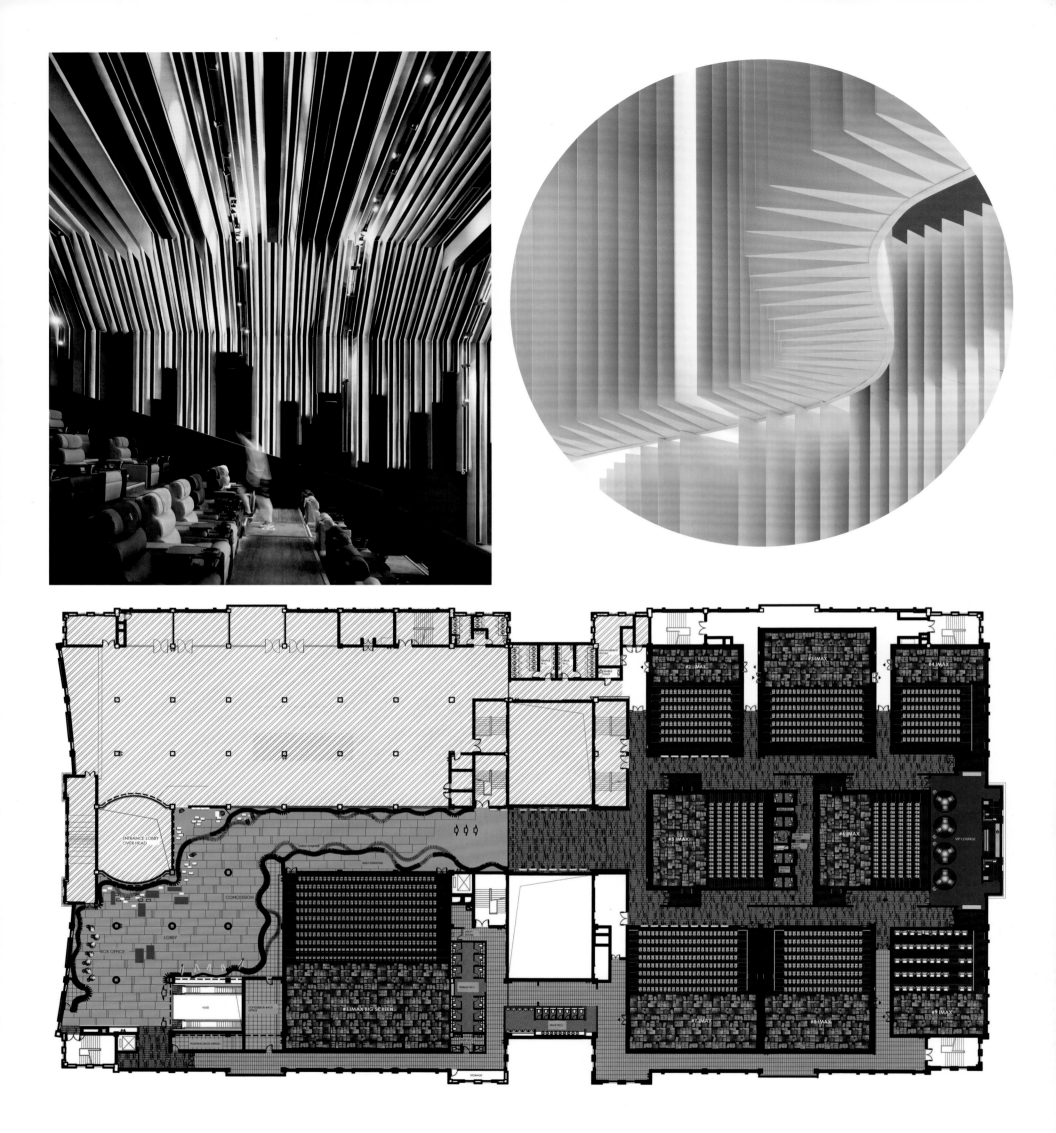

The cinema, located in a book city, has inspired the designers to combine the two elements into their design. Books are normally being published with white background and black fonts, whereas films, acting the opposite way, having images printed on black frames instead. The cinema uses this special correlation between books and films, by mix and matching their respective colors — white and black. First, you'd see pages of the books flipping along the wall. There're stacks of "paper" standing on the floor. When looking from afar, you will can't help but think they are real paper! In fact, they are the ticketing office made up of corian. Passing through the lobby into the hallway, the background switched from white to a striking black. Words of different fonts and colors scatter randomly against the black surface, forming famous dialogues extracted from classical movies.

这间电影院坐落在书城中,启发了设计师将书的元素融入设计。书本通常是白底黑字,电影却反其道而行,在黑色的胶卷上幻化出无穷的影像。本案的设计利用书籍和电影之间这种独特的关联,将代表性的黑白两色混合搭配,作为空间主色调。进入空间,书页仿佛沿着墙面翻动着,地板上有一些由书页堆叠起来的装置,整齐地摆放着,会让你禁不住觉得,它们是真实的纸张。而实际上,这是由人造石堆砌而成的售票处。穿过大堂来到走道,空间背景从白色切换到黑色。不同字体和颜色的单词,随机地散落,与黑色的表层形成对比,组合成经典电影中有名的词句。

● LIVING SPACE　　GOLD MEDAL
居住空间　　　　 金奖

D.R.HOME

设计共和·家

Design Agency / Neri&Hu Design and Research Office

D.R.home (Design Republic home) is located on the third floor of Design Republic Design Commune. The expression of design here is very casual yet modern, where basic functions are met but there's a lot of personal styling with some old objects and furniture. The public space is very loft-like, with openness that allows guests to move freely from one space to another. Smaller, more private corners offer settings that allow people to sit by themselves for more private moments. The bare space without the furniture was designed with flexibility in mind so that people could incorporate different setting in there in the future. The most unique space is the personal areas of closet/bathroom and also bedroom. Designers have combined the bathroom functions within a "storage concept" in the closet space.

Design Team / Lyndon Neri, Rossana Hu and others
Location / China
Area / 200 m²
Client / Design Republic

设计共和之家位于设计共和社区三楼。室内设计风格休闲而现代,满足空间基本功能的同时,装饰了许多具有个人风格的老物件和家具。公共空间非常像复式空间,宽敞开放,客人可以在各个空间中自由活动。小一些的更为私密的角落可供人们静坐,享受片刻私人空间。考虑到灵活性,空间中没有放置家具,因此未来可根据需要来配置不同的物件。最为特别的空间是私人区域,衣橱/浴室和卧室。设计师将浴室功能结合"储存概念",设置在衣橱空间中。

● SAMPLE SPACE　　GOLD MEDAL
样板房空间　　　　金奖

MADAM CHANEL

香奈儿女士

Design Agency / Chains Interior Ltd.

Madam Chanel is trendy and modern. She would like to have a space for weekend gathering.

By using glasses, metal, black and white color to show neat and elegant new urban ladies characters. Hexagonal puzzle screens divide the spaces and black and white balanced the space weightiness.

Black and white glasses not only create the light gradation but movements and reveal Madam Chanel's charm. ●

Location / Taiwan, China
Area / 300 m²
Client / Madam Chanel

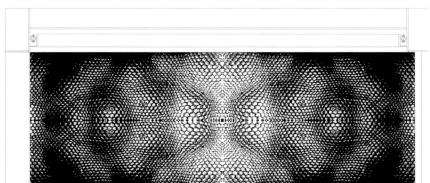

香奈儿女士喜爱时尚独有韵味。她想要一个在假日供朋友相聚的场所。

以玻璃、金属、黑白色调融合经典语汇线条呈现都会新女性特质。公共空间以六角隔屏界定场域关系，前后黑、白流动墙面平衡空间重量感。

黑白玻璃创造出光的渐变和移动，流露出香奈儿女士的迷人风采。

INSTALLATION & EXHIBITION SPACE	GOLD MEDAL
设施 / 展览空间	金奖

Design Team / Kris Lin, Yang Jiayu
Location / China
Area / 1,000 m²
Client / China Resources Land Limited

CITY CROSSING

华润中心

Design Agency / Kris Lin Interior Design (KLID)

In this project, designers undertake three parts of design work: architectural, interior and landscape.

This is the sales office of Huarun Center. The site is a slope and the road is higher while the project site is lower. There is 3 meters' height difference between the road and the site. There is a park at the other side of the road which assumes to be the most important interior landscape and also determines the building layout.

In order to have the view of the natural park, the groundwork of the sales office should be lifted to be higher than the floor level and it seems that the sales office building is floating on the water. A bridge is designed to connect the road and entrance of the sales office with which people can enter the interior space and approach the building while seeing the model through a horizontal glass window.

在这个项目中，设计师同时设计了建筑、室内、景观三个部分。

这是华润中心的售楼处，该处是一个斜坡，马路是高点，而项目所在位置是低点，项目与马路有三米的落差，而马路对面是一个自然公园，是室内最重要的景观面，进而也确定了建筑的布局与朝向。

为了能看到马路对面的自然公园，要将售楼处的主体架高，脱离地面，使整个售楼处主体看上去像是漂浮在水面上。项目设计了一座连接马路和售楼处入口的桥梁，人们接近建筑物和进入售楼处内部时可透过水平玻璃窗看到该模型。

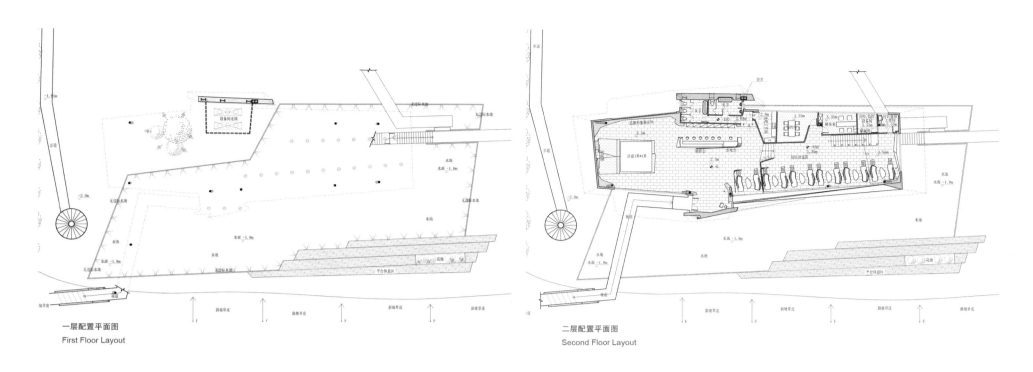

一层配置平面图
First Floor Layout

二层配置平面图
Second Floor Layout

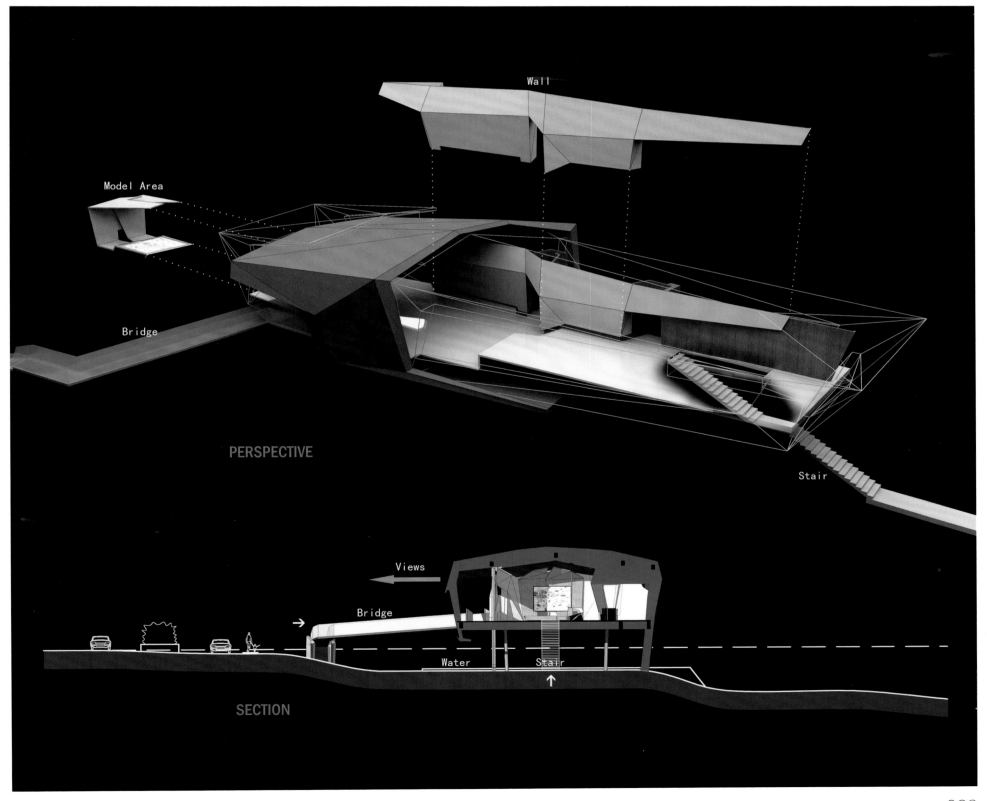

| PUBLIC SPACE | GOLD MEDAL |
| 公共空间 | 金奖 |

Design Team / Li Hu, Huang Wenjing, Qi Zhengdong, Thomas Batzenschlager, John Lim, Zhao Yao, Wang Jianling, Wu Lan, Ge Ruishi, Sigmund Lerner
Location / Qinhuangdao, China
Area / 2,700 m²
Client / Beijing Gehua Cultural Development Group, Initiate Development for Education and Service

GEHUA YOUTH AND CULTURAL CENTER

歌华营地体验中心

Design Agency / OPEN Architecture

This is a public welfare youth camp. Within the precious and very tight piece of land, we tried to maximize both the preservation of nature on site and the diversity of spatial qualities. Free flowing indoor spaces fully connect to the outdoor landscape. The same space can assume different functions for different occasions. The central courtyard is not only part of the landscape throughout the year but it is also an extension to the theatre for hosting a much larger crowd watching performances. When both sets of folding doors behind the stage are fully opened, the courtyard suddenly becomes a part of the theatre, creating a delightful surprise that transforms the small theatre into a large performing arena.

这是一个公益性的青少年营地，在紧张而宝贵的地块上，设计师尝试利用最少的资源去创造最大化、最丰富的空间体验，并且最大限度地保留自然。室内空间通透开放，不仅使阳光和风可以自在地穿过，也可灵活地适应不同的活动需求。建筑中心的内庭院，不仅是全年的景观，同时也可以扩展为观众席来观看剧场的演出。舞台后有两层大型折叠门，可以分别或同时打开，将室外庭院纳入剧场空间，创造令人惊喜的观演体验。

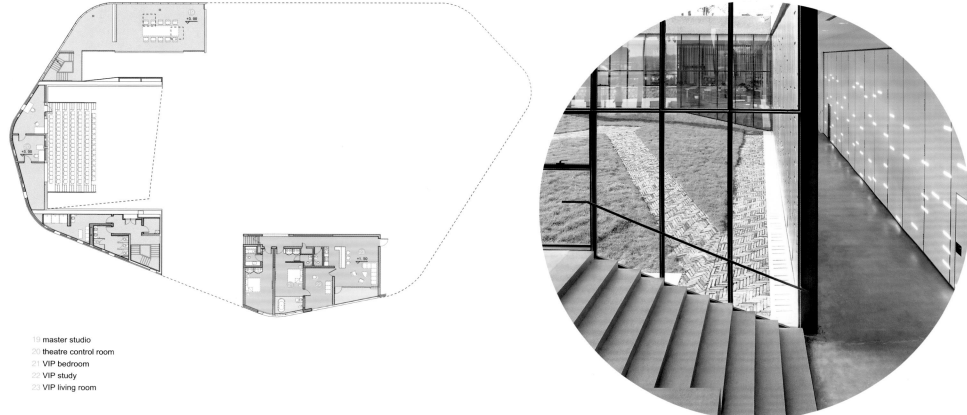

19 master studio
20 theatre control room
21 VIP bedroom
22 VIP study
23 VIP living room

● WORK SPACE
办公空间

GOLD MEDAL
金奖

Design Director / Project Director / Simon Park
Senior Concept Designer / Thomas Danet
Interior Designer / Jennifer Jiang
Construction Director / Jerry Wang
Project Manager / West Lu
Client Manager / Jack Zhang
Location / China
Area / 14,000 m²
Client / Adidas

ADIDAS GREATER CHINA HEADQUARTERS

阿迪达斯大中华区总部

Design Agency / PDM International

HEADQUARTERS

A revolutionary, "athletic-inspired" headquarters sporting the Adidas values.

The client facing areas were designed with a generic aesthetic as not to favour a certain brand line. However, the large showrooms dedicated to each brand were retail inspired and used as "mock-stores" for training.

Typical office space encompasses themed meeting spaces to express brand and image, along with a careful balance of unique hotspots providing collaboration areas on every floor to integrate teams and increase interaction, successfully capturing the overall design concept of the team player and Adidas's athletic spirit.

The central staircase binds all seven levels through an iconic "three stripe" lighting feature that penetrates all floors. A stunning cityscape forms the backdrop to the staircase and pantry delivering natural light, inspiring views and acting as a social hub and transitional space shared by all.

This cohesive workplace promotes a fun and engaging brand experience.

革命性的总部大楼承载着阿迪达斯的品牌价值。

对顾客开放区域的设计运用了通用的美学，并不单独突出某个特定的品牌线。然而，大大的样品陈列室则单独展示了每个品牌的零售理念，被当做"模拟店面"用于员工培训。

典型的办公空间包括展示了品牌形象的主题会议空间，精心平衡独特的热点，形成每层的协作区域，整合团队成员并增加员工间的互动。整体设计理念展现了良好的团队精神和阿迪达斯品牌的运动精神。

通过一个阿迪达斯标志性的"三条纹"照明装置，中央楼梯将这七层联系到一起。该照明可以穿透所有楼层。令人惊艳

的城市风光成为了楼梯和食品储藏室的大背景，这儿除自然光线外，更有怡人的风景。它还可作为社交枢纽，为所有人提供了一个过渡空间。

这个有凝聚力的工作场所提升了有趣且迷人的品牌体验。

HEADQUARTERS

● SHOPPING SPACE
购物空间

GOLD MEDAL
金奖

UMIX TOP UNDERWEAR MULTI-BRAND STORE

Umix 高端内衣综合品牌店

Design Agency / AS Design Service Limited

Design Team / Four Lau (Creative Director), Sam Sum (Art Director), Vincent Leung (Interior Designer)
Location / Macau, China
Area / 142 m²
Client / World First Holdings

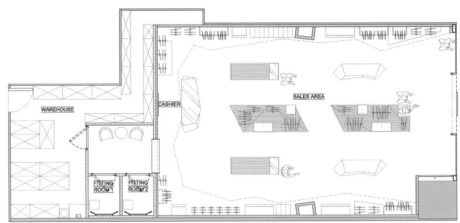

When two or more opinions and ideas occur simultaneously without causing contradictions, we call this compatibility phenomenon, coexistence, mutual tolerance and ideology compatibility.

UMix is a world famous brand shop in men and women underwear and casual wear. Designers feel there are many different opinions and philosophies between the sexes today, and hope to use the compatibility phenomenon as a new image, in order to achieve the core design concept of UMix shops, that men and women are compatible in the modern ideology of mutual tolerance.

The use of different size, different angles of polygons to describe their own opinions and the independence idea of both the sexes, and after being compatible, not only having idea of independence, but also it will create a more extended ideological significance and provide a new space for thinking philosophy.

当两种或多种理念同时出现而没有导致矛盾冲突，我们称之为兼容现象，又或者是共存。

UMix 是世界知名的经营高端男女内衣及休闲服装的时装店。设计师认为现今男女两性之间存在着许多不同的看法及理念，希望借由相容现象作为 UMix 专卖店的核心设计概念，以达到男女现代意识形态的兼容和共存。

运用不同尺寸和角度的多边形来形容男女两性各自的看法及自主理念，相容之后，不仅保有其各自的理念，同时还将创造出多个意识形态的延展，产生全新的思考空间。

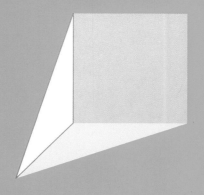

21st APIDA
FOOD SPACE
用餐空间

● FOOD SPACE
用餐空间

SILVER MEDAL
银奖

Design Team / Lyndon Neri, Rossana Hu and others
Location / China
Area / 76 m²
Client / Franck Pecol

Farine 面包房

Design Agency / Neri&Hu Design and Research Office

Aromatic coffee beans, sumptuous bread scent—Farine on Wukang Road is not just restauranteur Franck Pecol's new bakery in Shanghai, but also architects Neri&Hu's spatial interpretation of bakery as the progeny of coffee and bread. The design concept is encapsulated in where bread is made is a pure, white testing lab; where coffee is brewed, a textured palette of wood and raw metal reminiscent of the complex roasting process. The facade as a play of materials highlights and sometimes conceals the activities within, attracting visitors to double-take voyeuristically. Various scale and composition of aged bronze, reclaimed elm, textured glass, and even hints of colors also add to this layer of intrigue with their unique translucencies. What started as a pure lab space juxtaposing a raw, industrial coffee zone thus becomes a 2-in-1 shop where coffee and bread visually unite and compliment in taste. ●

芳香的咖啡豆，美好的面包香，位于武康路的 Farine 是业主 Franck Pecol 在上海新开张的面包房，由如恩设计研究室设计。这是一个胶囊状的纯白实验室，是烘焙面包的场所，是冲泡咖啡的地方，设计师运用了有纹理的木头和金属来让人联想到复杂的烘焙过程。外墙上利用材料打造出亮点，间或对店内的活动加以掩藏，吸引游客去了解更多。各种尺寸和构成的旧铜器，回收利用的榆木、玻璃，还有色彩等元素一起，增加了空间的层次。纯粹的实验室空间与原始工业感的咖啡区域合二为一形成店面空间，将咖啡和面包从视觉和味觉上结合到一起。

● FOOD SPACE
用餐空间

BRONZE MEDAL
铜奖

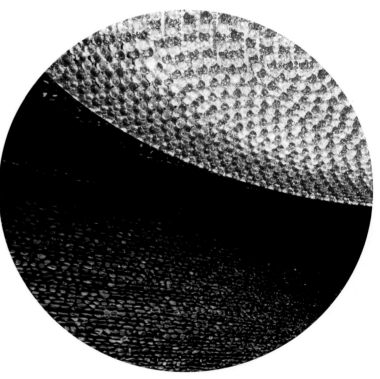

W HOTEL FEI ULTRA LOUNGE

W 酒店休息室

Design Agency / A.N.D.

Location / China
Area / 935 m²
Client / KWG Property Holdings Limited

Fei Ultra lounge is a bar lounge at W hotel in Guangzhou, China.

It is a large, cubic, glass-walled space with 18 meters each of three sides, and designers created two design concepts.

The first is to gently cover the enormous wall with a film of warm light, which is new in the world.

The color and flicker rate of the film, made up of countless glass fibers, can be controlled, and it constantly changes and sparkles.

It is succeeded in lighting up the whole wall with minimum lighting sources by a highly minute method of mixing the fiber and finishing gut.

The second is multi-layers of horizontal floors and vertical walls into that light-enveloped space.

It creates segmented individual rooms of different density and taste, welcomes guests and allows them to withdraw from their ordinary lives. They can experience astonishing time and space very only here. ●

本案是中国广州W酒店的休息吧。

这是一个由巨大的立体玻璃作为墙面的空间，三个侧面每侧都高达18米，设计师应用了两种设计理念。

首先是用温暖的光幕轻柔地覆盖在巨大的墙面上，此项设计乃世界首例。

由无数的玻璃纤维组成的光幕的颜色和闪烁频率，可以进行控制，不断变化闪烁。

这项设计成功地利用最低限度的照明资源点亮了整个墙面。

其次，在这个由光包覆着的空间中，设计了多层次的水平楼层和垂直墙面。

创造出分段的有着不同密度和品味的单独空间。休息室欢迎客人来此休憩，从日常生活中摆脱出来。在这里，人们可以感受到独一无二的迷人的时间和空间体验。

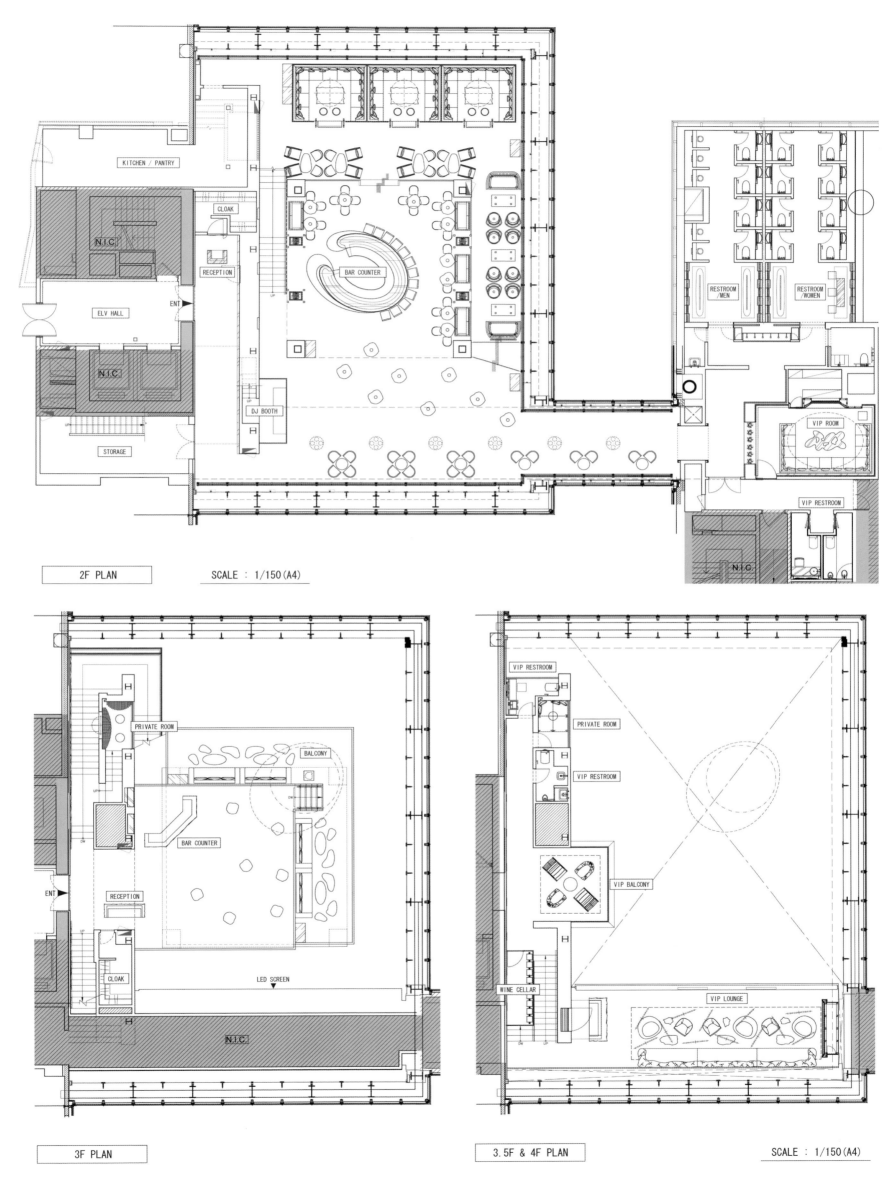

● FOOD SPACE　　EXCELLENCE MEDAL
用餐空间　　　　优秀奖

Design Team / Obayashi Corporation
Location / Japan
Area / 590 m²
Client / PALACE HOTEL

WADAKURA IN PALACE HOTEL

皇宫酒店 Wadakura 餐厅

Design Agency / A.N.D.

Considering both location and concept, Palace Hotel is the representative of hotels of Japan. Designers took in charge of the design of an 800-square-meter Japanese restaurant in the hotel. Since it is very difficult to create specialty in traditional Japanese-style space, they chose the basic concepts of design: traditional Japanese Decoration Method combined with High Criteria, and Newness.

The traditional Japanese buildings were usually built by wood and soil. Designers used the two materials separately to express the Newness. From the entrance, corridor to private rooms, they designed quiet spaces of soil by traditional plasterer craftsmen. After those spaces, there is a square with water features. Further to that, wooden yet modern spaces with privacy were created by splendid woodcrafter technique.

Designers initiated the new possibility of utilizing technique of craftsmen together with traditional materials, based on the lifestyle and culture of Japan. ●

考虑到项目的地理位置和理念，皇宫酒店堪称日本酒店的代表。设计师负责了该酒店中一间面积为800平方米的餐厅的设计。在传统日式风格的空间中创造出不一样的特色似乎有些不易，因此设计师采用了将传统日本装饰设计手法与高品质新颖时尚相结合的基本设计理念。

传统的日本建筑通常是用木材和泥土建造。设计师选择单独使用这两种材料来诠释新奇。从入口、走廊到私人空间，根据设计师的设计由传统泥水匠建造了一系列静谧空间。穿过这些空间，是一个设计了水景观的广场。此外，空间内还配备了一些木质的现代私密空间，利用杰出的木工技术打造而成。

在日本生活方式和文化的基础上，将工匠的精湛工艺与传统的材料相结合，创造出新的思路。

IN PALACE HOTEL

● FOOD SPACE
用餐空间

EXCELLENCE MEDAL
优秀奖

Design Team / Lai Siew Hong, Loo Shook Mung, Stanley Yap, Elaine Yap, Hanako Suzuka, Fang Fang Tan, Mak Sook Har
Location / Hong Kong, China
Area / 677 m²
Client / Sun Hung Kai Properties Ltd

JIN & INA

牛郎 & 织女

Design Agency / Blu Water Studio Sdn Bhd

The concept of Jin & Ina – Chinese and Japanese restaurants in Hong Kong was inspired by a forbidden love of the Princess & the Cowherd.

The fairytale known in China and Japan, was the starting point of the design vision and provided for the division of both restaurants into heaven and earth. Common elements found in both cultures helped to connect the design, while details typical of Chinese or Japanese heritage defined the spaces.

Braided knots, the symbol of neverending love, implemented on crystal screens welcome patrons. Upon entering, guests experience the earthly elements of natural timber and stone. The vibrant artwork displayed motifs of birds symbolizing the connection of earth with heaven like the bridge of magpies that unites the lovers. Custom crystal lighting that represents the heaven is displayed amongst the windows. Spaces are divided by a sliding partition, which can allow both restaurants to unite as one.

本案是位于香港的一间中日式餐厅，设计理念灵感来自于牛郎织女之间一段不被允许的爱恋。

这个在中国和日本都很有名的神话故事是设计的出发点，设计师将餐厅空间分为天堂和尘世两个部分。两种文化中都常见的元素将空间连接成一个整体，中式和日式的典型细节设计则定义了空间的特色。

象征着无止境的爱恋的编织结被运用在欢迎来客的水晶帘上。甫进空间，客人会看到天然的木材和石块等尘世常见的元素。绘制鸟类图案的充满生气的艺术作品代表着尘世与天堂之间的联系，正如喜鹊为恋人搭成的桥梁。定制的水晶灯代表了天堂。空间被滑动的隔断隔开，整个餐厅空间成为一个整体。

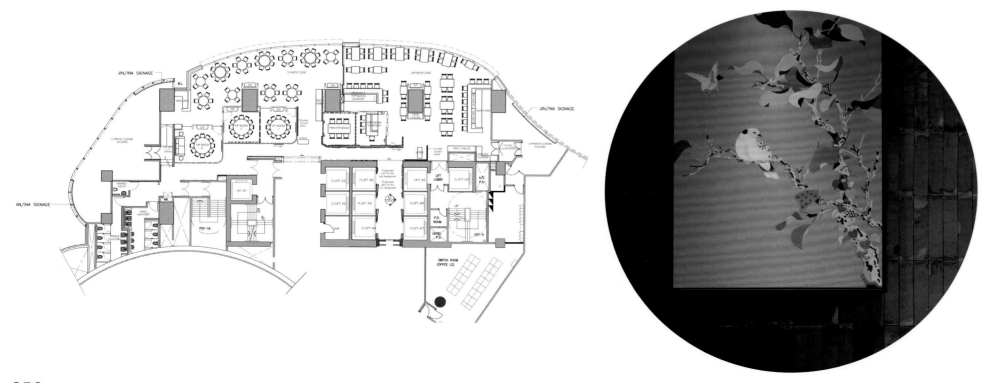

● FOOD SPACE
用餐空间

TOP TEN
十名入围

HAZEL & HERSHEY CAFÉ

Hazel & Hershey 咖啡

Design Agency / Atelier E Limited

Design Team / Enoch Hui, Nami Tsui
Location / Hong Kong, China
Area / 100 m²
Client / Mr Birdie Chiu

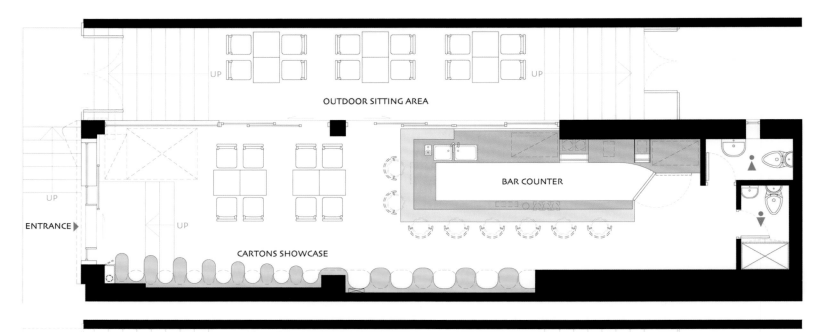
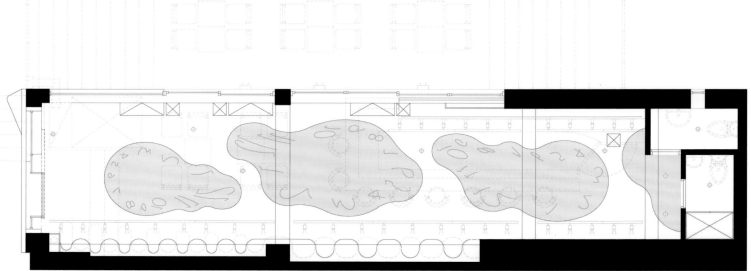

A coffee shop tucked in a narrow alley near the SOHO district which sells a full range of coffee makers.

Designers used a 2-tone graphics which resembles a coffee cup overflowing with hot, foamed milk to send a clear message to the passer-by; the graphical treatment is deliberately curved and smoothed to produce a sharp contrast to its surrounding texture.

The client is very environmentally conscious, the core component of this design was created using thousands of layers of recycled carton, each individually cut, to form an organic waveform that becomes a product display wall stretching the entire length of the shop. Over time it would absorb the smell of coffee to become part of the aroma of the interior.

Finally, 3 needleless melting clocks decorate the ceiling to remind the patrons that the best coffee in town is not meant to be enjoyed in a hurry.

本案是一间邻近SOHO区，位于小巷子里的咖啡店，销售各种咖啡机。

设计师运用类似装满泡沫奶油，散发着热气的咖啡杯图形，给路过的人们传达清晰的信息，该平面图形经过特别处理来与周围的织物形成强烈反差。

客户非常有环保意识，设计的核心部件是利用上千块回收的纸板制作而成，每块纸板都被单独切割，最终形成有机的波浪型产品展示墙，贯穿整个店面。随着时间的流逝，纸板会吸收咖啡的香味，最终成为室内芳香的一部分。

最后，3个没有指针的金属钟装饰在天花板上，提醒顾客们，最好的咖啡注定要慢慢品尝才能更好地知其味。

● **FOOD SPACE**
用餐空间

TOP TEN
十名入围

ENJOY COFFEE

享入啡啡

Design Agency / Fuzhou Lianxu Decoration Design Co., Ltd.

This is a natural space without seditious design. A free and easy space derives from designer's in-depth knowledge about materials and strict standard about details. It is an ideal leisure place full of sunshine and fresh air. Designer blends ordinary materials in design, and transforms an ordinary building into an elegant comfortable and modern café with warm atmosphere. Wood panel with simple treatment is used on the wall, with its node and texture clearly visible. Use simple material combination which is beautiful and enriched details instead of common suspended ceiling materials, expressing the eco-friendly design theme. Warm tone and materials with proper decorations in the space, combining with natural aesthetics, create relaxed atmosphere. ●

Designer / Wu Lianxu
Location / China
Area / 480 m²

没有煽动性的颠覆设计，一切从自然出发。一个洒脱的空间源于设计师对材料的深入了解和对细节的严格追求，创造了一个充满阳光和新鲜空气的休闲好去处。将常见的材料融入设计，一个普通的建筑结构转换成一个具有温暖的氛围，美观舒适兼具现代感的咖啡厅。木板简单护理便素颜上墙，结节纹理毫不隐藏。简单的材质组合取代了常见的吊顶材料，美观的同时丰富了空间的局部，也表达了设计师的设计主张，环保是设计师的旋律。融入自然美学，温暖的材料、温暖的色调加入些许装饰，让人放松心情呼吸自然。

COFFEE

FOOD SPACE	TOP TEN			Designer / Stefano Tordiglione
用餐空间	十名入围			Location / Hong Kong, China
				Area / 300 m²
				Client / Woolly Pig Concepts

SAL CURIOSO

Sal Curioso 餐厅

Design Agency / Stefano Tordiglione Design Ltd

The concept of the design is based on the life personality of the imaginary food inventor and traveller Sal Curioso. The designer sliced the space through the middle to create a distinctive split effect centred around a construct and deconstruct concept.

Divided into distinct areas, the lounge is a welcoming cozy space, inspired by clubs of the 70s with optical wallpaper, sofas covered in colourful fabric and stand-out black and red stools. Quirky and bright design themes continue throughout. From the bold, cut Spanish tiles of the mosaic bar, in keeping with the overriding construct and deconstruct concept and inspired by Spanish architect Gaudi's decoration, to the mixed-era furniture including '50s chairs and '60s banquettes exclusively designed by the designer and '70s vintage lamps in the chef's table, the restaurant is a delightful gallimaufry of eras and worlds that come together under one roof.

餐厅的设计灵感来自于虚幻的美食发明家 Sal Curioso。设计师通过解构和重组的概念，创造出与众不同的空间。

酒吧区是一个温馨惬意的空间，摆放了色彩鲜艳的布沙发及鲜明的红黑色矮凳。马赛克酒吧台运用了分割的西班牙瓷砖，与建筑师高迪的绚丽装饰风格保持一致。餐厅中央的点状装饰图案能让人联想到西班牙的阿尔罕布拉宫。整个餐厅充满了南欧风情。

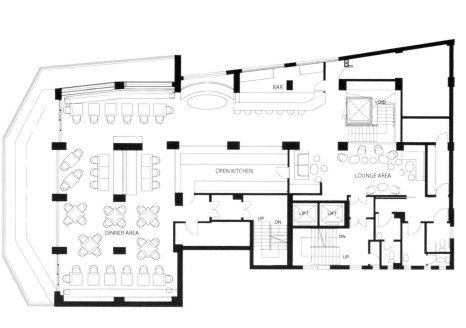

LAYOUT PLAN
SCALE 1:100

CURIOSO

069

- FOOD SPACE
  用餐空间
- TOP TEN
  十名入围

CAPO RESTAURANT

Capo 餐厅

Design Agency / Neri&Hu Design and Research Office

The Basilica, an architectural typology of a bygone era that is all too often reduced to a page in the history book or a stop on the tourist route, is revisited as a design concept for a modern Italian restaurant by Neri&Hu. Traditionally at the heart of civic activity, the Basilica was once a bustling public space—a church, a courthouse, a stage, a gallery, and a dining hall all at once—a place to see and be seen. Neri&Hu's design intent for Capo was to embody the spirit of the Basilica within this cavernous space carved out of the attic of a 1911 building located on the historic Bund of Shanghai. Rather than default to an open plan configuration, Neri&Hu makes the conscious decision to carve up the restaurant into a series of distinct rooms at various scales, each with a unique character.

Design Team / Lyndon Neri, Rossana Hu and others
Location / China
Area / 600 m²
Client / JIA Boutique Hotels

Basilica（长方形会堂），是过去的一种建筑类型，现在通常只是出现在历史课本中或是观光旅行地。如恩设计将其转变成一种设计理念，应用在一间现代的意大利餐厅中。过去作为市民活动的中心，Basilica 也曾经是熙熙攘攘的公共场所——教堂、法庭、舞台、美术馆和食堂等。如恩设计对于卡波餐厅的设计目标是，在这栋位于上海的 1911 年的建筑物顶层的洞穴状空间中体现 Basilica 的精神。如恩设计将餐厅划分成一系列大小各异的空间，每个空间都有其独特的特征。

FOOD SPACE	TOP TEN			Design Team / Lai Siew Hong, Hanako Suzuka, Kenny Tan, Mak Sook Har

Design Team / Lai Siew Hong, Hanako Suzuka, Kenny Tan, Mak Sook Har
Location / Malaysia
Area / 378 m²
Client / Kampachi Restaurants Sdn Bhd

KAMPACHI AT THE TROIKA

Kampachi 餐厅

Design Agency / Blu Water Studio Sdn Bhd

The comfort of an open fire, soothing sounds of crackling firewood and richness of timber epitomizes the ambiance of a traditional Japanese hearth. Together with the reminiscent feel of coarse stone and roughened iron cookware, these elements set the atmosphere for the new Kampachi at Troika, Kuala Lumpur, iconic restaurant by Equatorial specializing in sophisticated Japanese cuisine.

Designers united contemporary modernity with the nostalgic past by maintaining the existing concrete elements together with the irregular lines of the building and complementing it with materials recycled from the Equatorial hotel, where 40 years ago the first Kampachi restaurant was opened.

The feature screen presents a photo taken by the owner during a trip to Japan providing a fragmented view of a temple rooftop. While looking above, a cloud of customized glass pendants twinkles throughout the void space and reflects on the mirrored louver ceiling creating a magical overall impression. ●

舒适的火光，不时发出噼啪声响的柴火，大量的木材，这是对传统日式灶台氛围的一个概括。设计师运用粗糙的石块和铁质厨具等元素在 Kampachi 餐厅中创造出怀旧的氛围。

设计师将现代时尚与旧日情怀相结合，保留了现有的混凝土元素及建筑物的不规则线条，用一些回收自 Equatorial 酒店（40 年前第一间 Kampachi 餐厅所在地）的材料对空间进行了补充。

特色屏风上是一幅业主在日本旅游时拍摄的照片。朝顶上望去，特别定制的云状玻璃吊灯在空间中闪烁，反照在天窗玻璃上，创造出魔幻般的整体氛围。

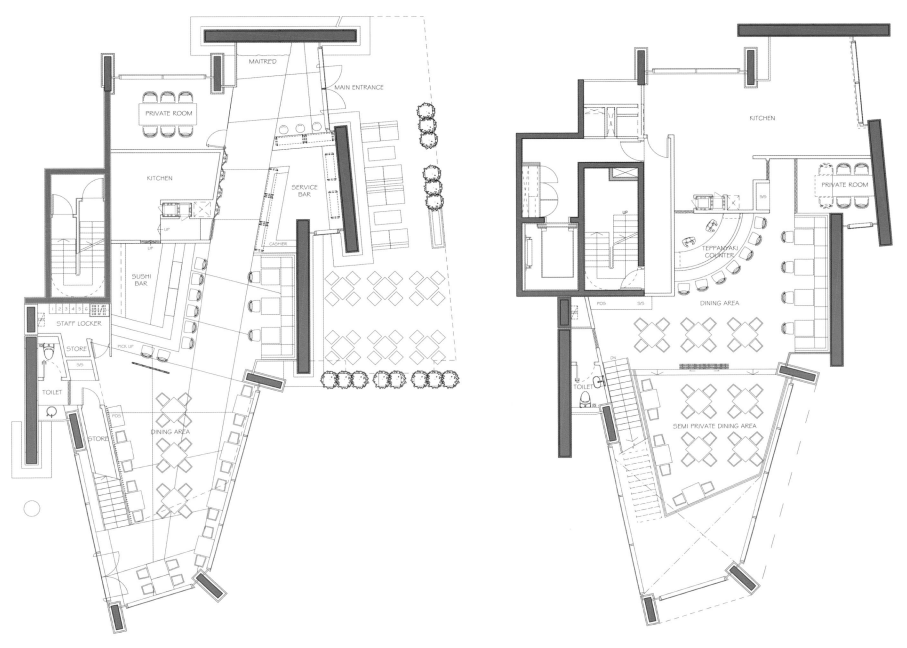

21st APIDA

HOTEL SPACE
酒店空间

| HOTEL SPACE 酒店空间 | SILVER MEDAL 银奖 | | Design Team / Clint Nagata, Kraisupa Asvinvichit Location / Thailand Area / 70,400 m² Client / Pacific Phuket |

REGENT PHUKET CAPE PANWA

普吉岛攀瓦角丽晶酒店

Design Agency / BLINK Design Group Pte Ltd

The design signified the revamp of one of the most renowned Asian hospitality brands, The Regent, featuring 70 rooms with sublime panoramic sea views and 35 pool suites.

Inspired by Sino-Portuguese heritage, furniture is custom-designed with key featured pieces selected from bespoke works by Thai artists with a nod to Phuket's past pearl and tin-mining industries, including sea pots, chinaware, and some wall-hung pieces. The materials used are all natural, including silk, jute rugs, rattan, bamboo and teak wood, some of which are sourced from the Mae Fah Luang Foundation, a Thai non-profit organization founded in 1972 that preserves the environment, and supporting local art and traditional culture.

The lobby's interior architecture features soaring ceilings that reveal the exposed roof structure of the classic Thai arch, finished with natural rattan and the characteristic Thai-green accent colour as well as bespoke classic artworks.

本案是对亚洲知名酒店品牌丽晶酒店的一次重装设计。该酒店有70间全海景房,35间泳池套房。

设计灵感来自于中葡传统文化,家具是特别定制的,主要特色物件是从泰国艺术家的定制作品中精心挑选出来的,向普吉岛过去的珍珠及采矿工业致敬,包括海锅子、陶瓷器,以及一些挂在墙上的物件。采用了天然的材料,包括丝织物、黄麻纤维毯、藤条、竹子和柚木,其中一些来自于皇太后基金会——泰国一家成立于1972年的非盈利性组织,致力于环境保护,为当地艺术和传统文化提供支持。

大堂室内建筑的特色是高耸的天花板,经典的泰式建筑屋顶结构一览无遗,饰面材料则使用了天然的藤条和极具特色的泰国绿,另外还有定制的经典艺术作品。

CAPE PANWA

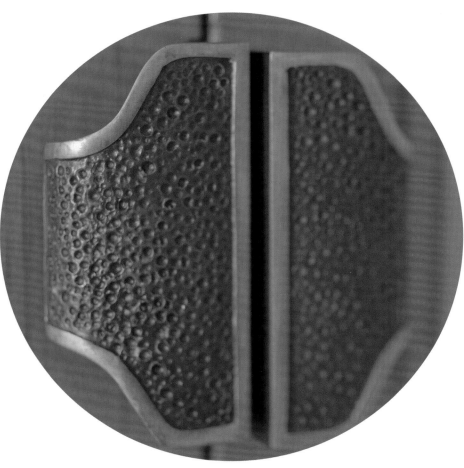

HOTEL SPACE
酒店空间

BRONZE MEDAL
铜奖

Design Team / Ben Diu, Charlotte Varley, David Wincey, Florrie Chow, Francesco Sacconi, Ivan Li, Jack Yeung, Mazen El Mahmoud, Stephen Jones
Location / Hong Kong, China
Area / 2,700 m²
Client / InterContinental Hotels Group

INTERCONTINENTAL HONG KONG (BALLROOM, LOBBY AND LOBBY LOUNGE)

香港洲际酒店（宴会厅、大堂和大堂酒廊）

Design Agency / Woods Bagot

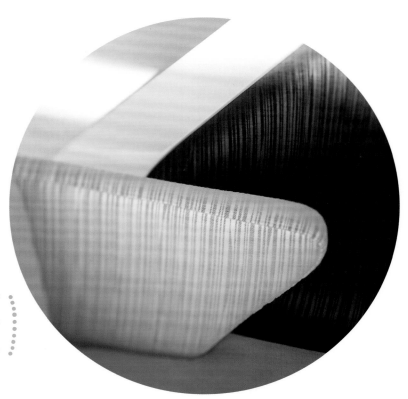

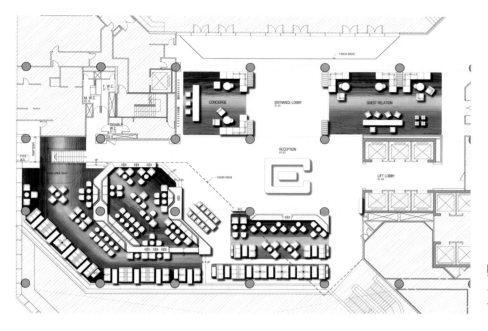
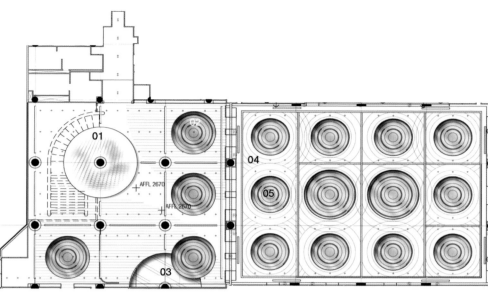

The lobby lounge offers the most extraordinary vantage point from which to experience Hong Kong. Its wrap-around, three-storey plate glass windows and panoramic harbour views have made this venue the place to see and be seen for over three decades. The new design needed to be a subtle enhancement to the existing facility rather than an overpowering new design that would compete with the iconic views.

The client's vision for the new Ballroom design was both glamorous and contemporary — understated luxury that would appeal to the various market sectors and align successfully with the overall brand of the InterContinental Hotel Hong Kong with its refined Asian luxury brand. When designing both the ballroom and the lobby lounge, sustainable initiatives were considered from the early concept phase, with locally sourced materials also utilised for both spaces. ●

大堂酒廊所处位置拥有极佳的视角。空间进行了环绕式处理，拥有三层高的落地窗，能欣赏到维多利亚港的全景。新的设计微妙地增强了原有的设施，进一步完善了其形象特征。

客户对于新宴会厅的设计要求是迷人而现代，以低调的奢华来吸引多层次的顾客，同时与洲际酒店这一奢华酒店品牌的整体形象相符合。在设计宴会厅及大堂酒廊的最初阶段，考虑到了可持续发展的诉求，选用了当地的材料。

(BALLROOM, LOBBY AND LOBBY LOUNGE)

(BALLROOM, LOBBY AND LOBBY LOUNGE)

● HOTEL SPACE
酒店空间

EXCELLENCE MEDAL
优秀奖

MANDARIN ORIENTAL PUDONG SHANGHAI

上海浦东文华东方酒店

Design Agency / Jiang & Associates Design

First of all, quantities of succinct lines as well as smooth style furniture are used to intensify modern atmosphere in the whole space; secondly, the design metaphorically embodies the Huangpu River — which has fairly strong Shanghai characteristics — plus screens and glass panes, leads guests to inadvertently find out the classic orient romance hidden in up-to-date modernity; moreover, exaggerated patterns and exquisite artworks are applied to create elegant artistic ambience; lastly, the designer gives up dense colors and uniform tones to adopt colorful and pastel translucent materials in abundance to broaden the long and narrow public space. ●

Design Team / Frank Jiang, J Lee Rofkind
Location / China
Area / 66,000 m²
Client / Shanghai Ruiming Real Estate Co., Ltd.

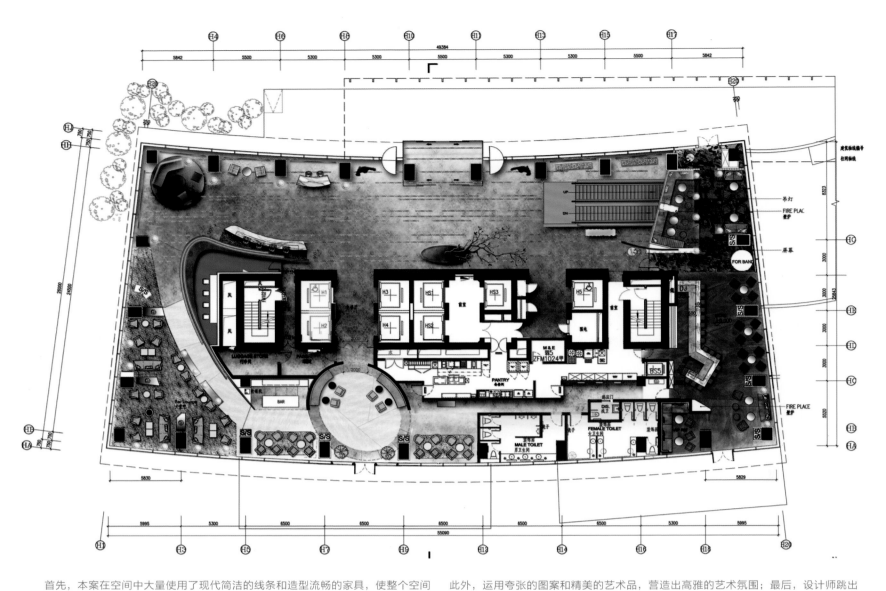

首先，本案在空间中大量使用了现代简洁的线条和造型流畅的家具，使整个空间充满现代气息；其次，用隐喻的手法将具有上海特色的元素黄浦江，以及屏风和窗格等元素融入整个设计中，让客人在不经意间发现现代风格中隐藏的古典东方的浪漫；此外，运用夸张的图案和精美的艺术品，营造出高雅的艺术氛围；最后，设计师跳出浓重的色彩和统一的色调，大量采用缤纷柔和的半透明材质，使原本长而窄的公共空间变得通透而开阔。

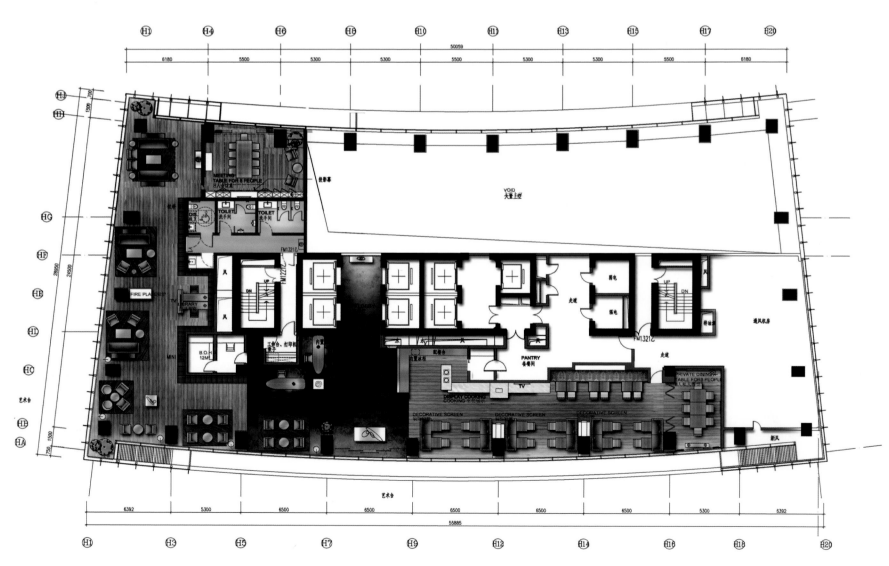

098　MANDARIN ORIENTAL

PUDONG SHANGHAI

● HOTEL SPACE
酒店空间

EXCELLENCE MEDAL
优秀奖

HAITANG BAY NO. 9 RESORT SANYA

三亚海棠湾 9 号度假酒店

Design Agency / Yangbangsheng & Associates Group

Design Team / Yang Bangsheng, Huang Shengguang
Location / China
Area / 80,000 m²
Client / Chinese PLA General Hospital

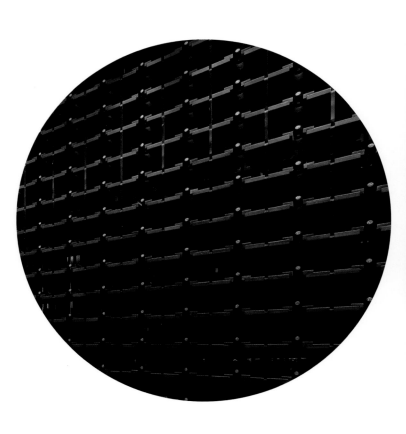

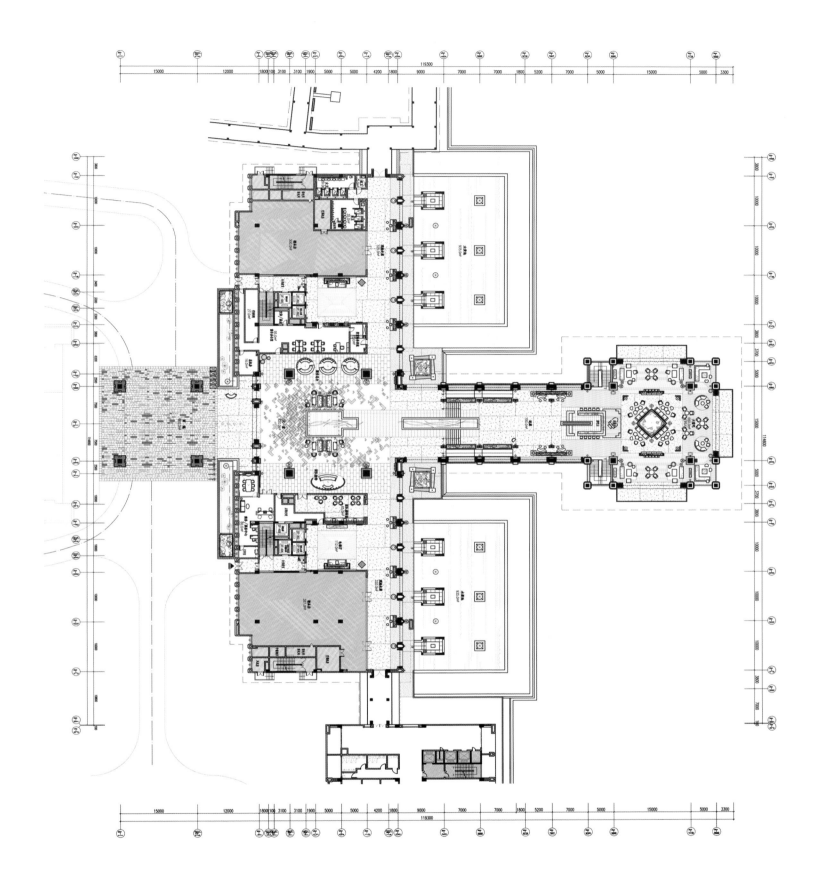

The design of this project combines modern Chinese style and tropical island experience. Pinnacle form lobby line up with lobby bar, highlights the grand and magnificent hotel. Wood ceiling, Chinese style screen and modern Chinese style furniture express meaningful Chinese charm. In the public area, open folding door replaced exterior wall, blend the inside and outside space together. Decoration pot, thatch pitched roof and brocade of Li nationality pattern of Southeast Asia characteristic, create a natural and romantic space for vacation.

该项目设计融合了现代中式风格与热带海岛风情。尖顶造型的大堂与大堂吧连成一线，凸显酒店的正气与恢宏。木结构天花、中式屏风以及现代中式家具的搭配传递出隽永的中式韵味。公共区域内开敞的折叠门取代外墙，将室内外空间融为一体。东南亚特色的装饰罐、茅草斜屋顶以及黎锦纹样的装饰图案，共同营造出自然浪漫的度假空间。

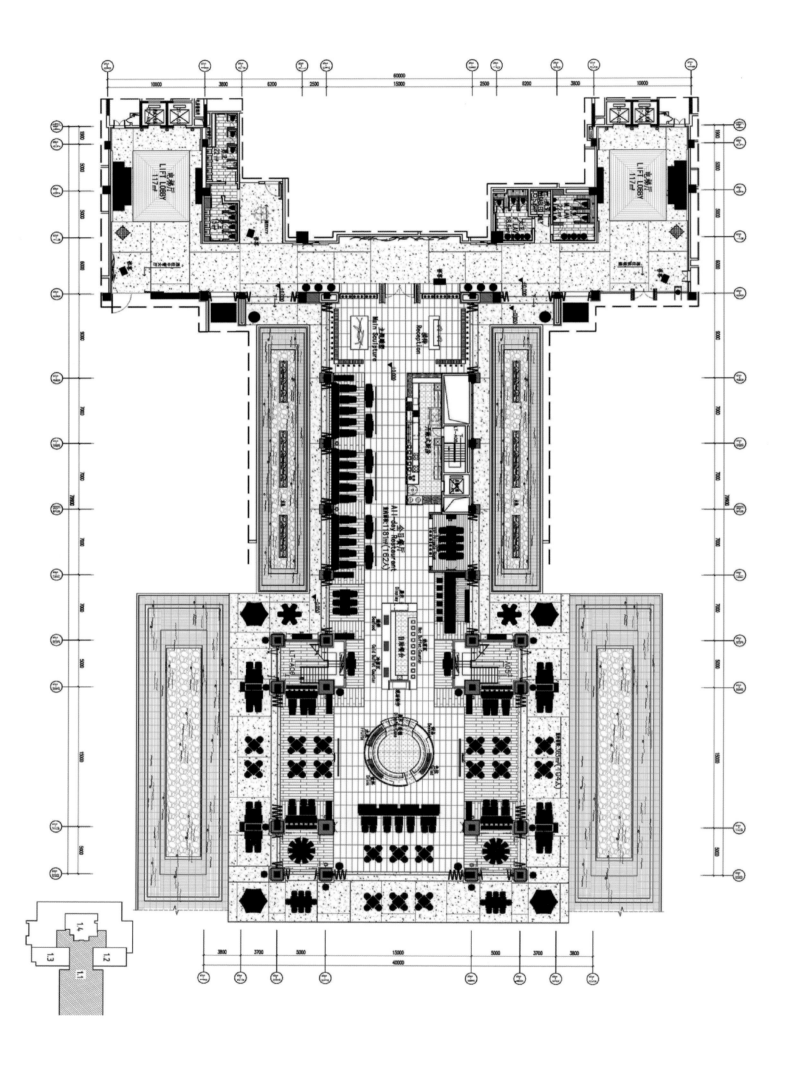

104 HAITANG BAY NO. 9

● HOTEL SPACE 酒店空间   TOP TEN 十名入围

HOLIDAY INN RESORT, CHANGBAI MOUNTAIN

长白山万达假日度假酒店

Design Agency / BLVD

The four star resort hotel sits comfortably nestled at the very base of the majestic mountain making it an extremely convenient base for the multitude of winter time skiers and summer time mountain trekkers that congregate for leisure. The ski chalet style architecture is laid out as a three pronged arrangement with the ground floor Main Entrance and Reception Lobby block at the center flanked on three sides, seemingly wrapping around the base of the mountain, by the Restaurant and Ski Bar, Spa and indoor Swimming Pool and the Banquet and Conference facilities.

With the natural wondrous environment in mind, the interior design concept uses uncomplicated and natural finishes such as wood paneled walls and columns, wood strip flooring, rough and polished stone, off-white plaster in a near natural state, timber and wicker furniture with leather and single tone upholstery and fabrics.

Designer/ Liu Honglei
Location / China
Area / 40,000 m²
Client / Changbai Mountain International Tourism Resort Development Co., Ltd.

该四星级度假酒店舒适地坐落在雄伟的高山中,为众多冬季滑雪者和夏季登山者来度假休闲提供了一个极其便利的场所。首层主入口和接待大厅位于中心,滑雪小屋风格的建筑则从三侧将其包围,看起来像是餐厅和滑雪酒吧、水疗中心、室内游泳池、宴会厅及会议室将山底包裹起来。

考虑到其周围优美的自然环境,室内设计使用了简单和自然的表层材料,例如墙面和柱子上的木饰板、木条地板、原始抛光的石块、自然形态的灰白色石膏、饰有皮革和软垫的木制和柳条编织家具以及各种织物。

1 LOBBY
2 LOBBY BAR
3 KITCHEN
4 ALL DAY DINING
5 SKI BAR
6 RESORT CENTER
7 GYM
8 CHANGE ROOMS
9 SWIMMING POOL
10 SPA TREATMENT ROOM
11 FOOT MASSAGE ROOM
12 KIDS GAME ROOM
13 KITCHEN
14 VIP ROOM
15 CLOAKROOM
16 BALLROOM
17 BALLROOM FOYER

Holiday Inn Resort Changbai Shan Level 1-Floor Plan

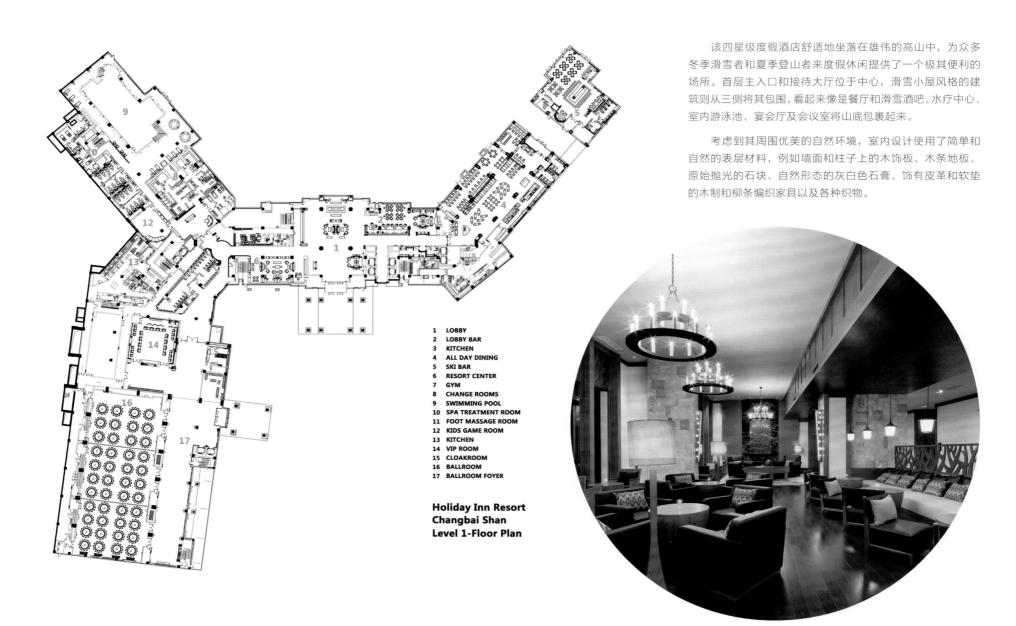

- **HOTEL SPACE** 酒店空间
- **TOP TEN** 十名入围

时尚精品饭店会所酒店

Design Agency / L'atelier Fantasia

This is contemporary art style space with clean lines and proper color scheme. Tableware with similar color scheme echoes with furniture and the space, reflecting the good taste of the owner. Furniture with brief shape blends into the space, creating stylish and concise overall atmosphere. Open type kitchen of dark color scheme and the suspended sphere lamp in the dining room create party-like romantic ambience, during cooking and dining.

　　本案是当代艺术风格空间，拥有利落线条和合理的色彩构成。相似色系的餐具、家具和空间呼应，进而展现主人对于布置的用心及好品味。家具以当代简约线条融入空间，为整体添加时尚简洁氛围，深色系的开放式厨房、垂吊式的球形餐厅灯具，令烹饪到用餐，都仿佛具有派对中身历其境的浪漫。

Design Team / IDAN CHIANG, SAM WU
Location / Taiwan, China
Area / 226 m²
Client / MR CHOU

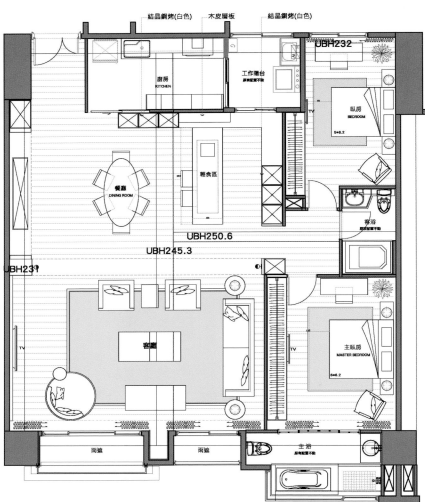

■ HOTEL SPACE　　　TOP TEN
酒店空间　　　　　十名入围

HUBIN SPRING SEASON HOTEL

湖滨四季春酒店

Design Agency / S-ZONA DESIGNER CONSULTANT CO.,LTD

The location of this project is nearby the lake and the roads which can give us an enjoyment of natural beauty, and at the same time ensures us the convenience of transportation. The significant and solemn yet simple Euro-style architecture together with the spacious and extraordinary topic hotel, they both provide the basis for the positioning of space design — full and rich, with strong sense of dimension and striking sense of seasons.

In the execution of this design, through obvious color system with strong visual impact, the four seasons impression of Jiangnan under the interpretation of international tactics and various types of furnishings, a passionate yet noble refined public house with the transitional characteristics of amalgamating Chinese and Western cultures are built.

Design Team / Feng Jiayun, Fan Riqiao, Sun Liming, Guo Xufeng
Location / China
Area / 9,000 m²
Client / Wuxi Hubin Spring Season Hotel Co., Ltd.

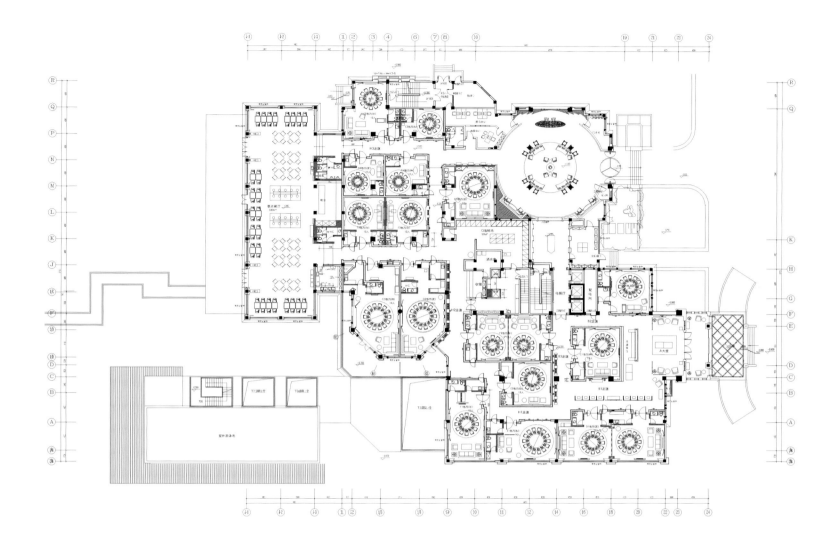

二层平面布置图

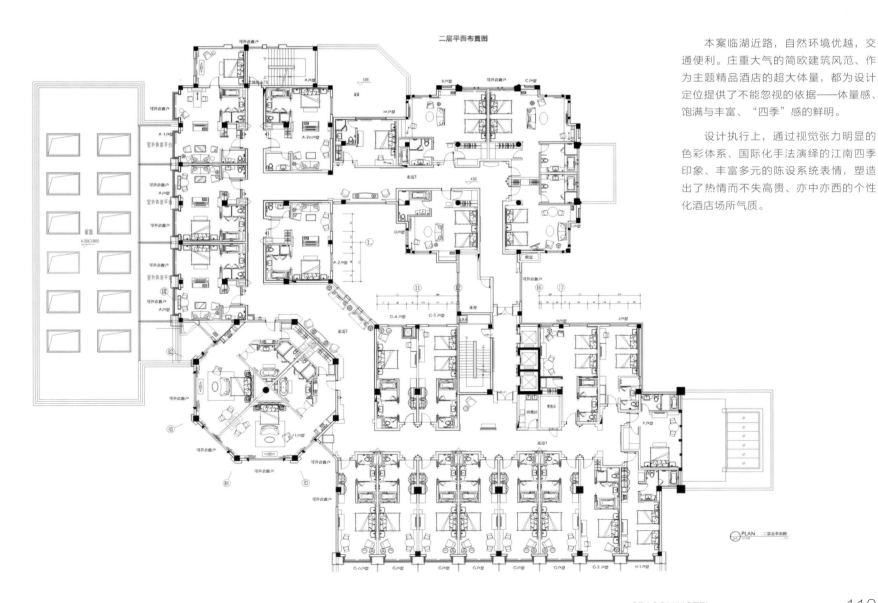

本案临湖近路，自然环境优越，交通便利。庄重大气的简欧建筑风范、作为主题精品酒店的超大体量，都为设计定位提供了不能忽视的依据——体量感、饱满与丰富、"四季"感的鲜明。

设计执行上，通过视觉张力明显的色彩体系、国际化手法演绎的江南四季印象、丰富多元的陈设系统表情，塑造出了热情而不失高贵、亦中亦西的个性化酒店场所气质。

SEASON HOTEL

● HOTEL SPACE TOP TEN
  酒店空间   十名入围

WUHAN LIGHT TEXTILE EMPLOYEE SANATORIUM 4TH PHASE — THOUSANDS STREAM TO THE MANSION

武汉市轻纺职工疗养院 4 期·千水归堂

Design Agency / Wuhan Yinhan Art Engineering Co., Ltd.

Designer / Chen Lei
Location / China
Area / 5,400 m²
Client / Wuhan employee sanatorium infrastructure construction department

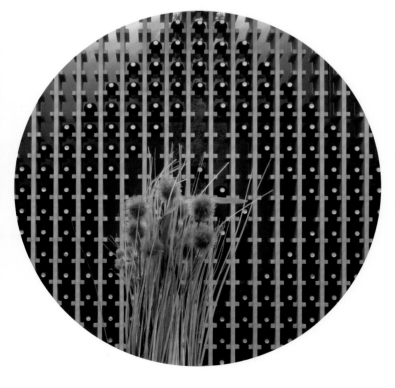

Seclusion at the forest, rest at the place that surrounded by water. This is the most characteristic geographical environment of this project. Design target of the project is to create soft and quiet space, and more importantly, to express the beauty of nature through design. Raw eco-friendly materials such as wood, stone and linen naturally make a dialogue with nature. There are 13 courtyards to make sure of enough natural light and ventilation in every single corner of the building. The core courtyard of the building is created to retain two camphor trees. Let these trees continue to narrate their stories.

归隐于林，栖于四面环水之地，这是此项目最具特色的地理环境。设计理念不仅追求柔和、安静，更重要的是让设计来衬托大自然的美好。原木、石材、亚麻布，这些原汁原味的生态材料与自然进行对话。室内分布着 13 个天井，确保建筑每一个角落的采光和通风。为保留两棵大樟树而形成建筑的核心天井。让树木继续叙述它们的故事。

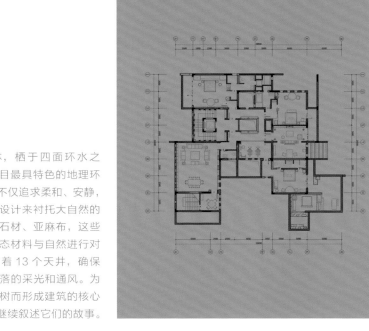
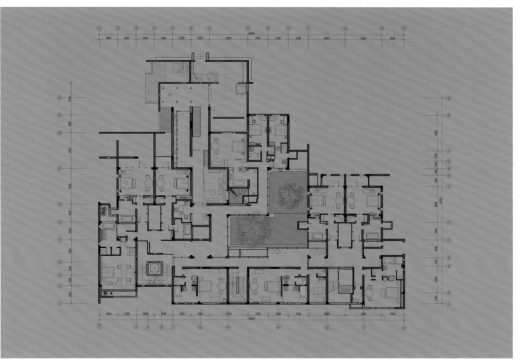

HOTEL SPACE
酒店空间

TOP TEN
十名入围

ALOFT KL SENTRAL

中央车站雅乐轩酒店

Design Agency / Blu Water Studio Sdn Bhd

Urban design mixed with local influences defines the style for Aloft KL Sentral.

As the biggest Aloft in the world, hotel accommodates the grand ballroom inspired by the walk in the park. Oversized pendants correspond with the hibiscus and leaves carpet pattern. The modern touch comes from the profilit glass and recycled wood fibre wall panels.

The high-ceiling rooms were inspired by the mood of industrial lofts. To liven up the space the brand signature design palettes were mixed with vibrant hues. As the focal point every room features the headboard with color prints of cartoons by artist Antares. The illustrations translate the Malaysian lifestyle with a local sense of humor. The artwork is punctuated with natural materials and textures including cork and wood to warm up the loft feel. Also the dramatic room corridors in grey were highlighted by the vivid paintings by local renowned artist Yusuf Gajah.

Design Team / Lai Siew Hong, Lee Ping Seng, Kenny Tan, Mak Sook Har
Location / Malaysia
Area / 22,411 m²
Client / Iringan Flora Sdn Bhd (hotel operator Starwood Hotels and Resorts)

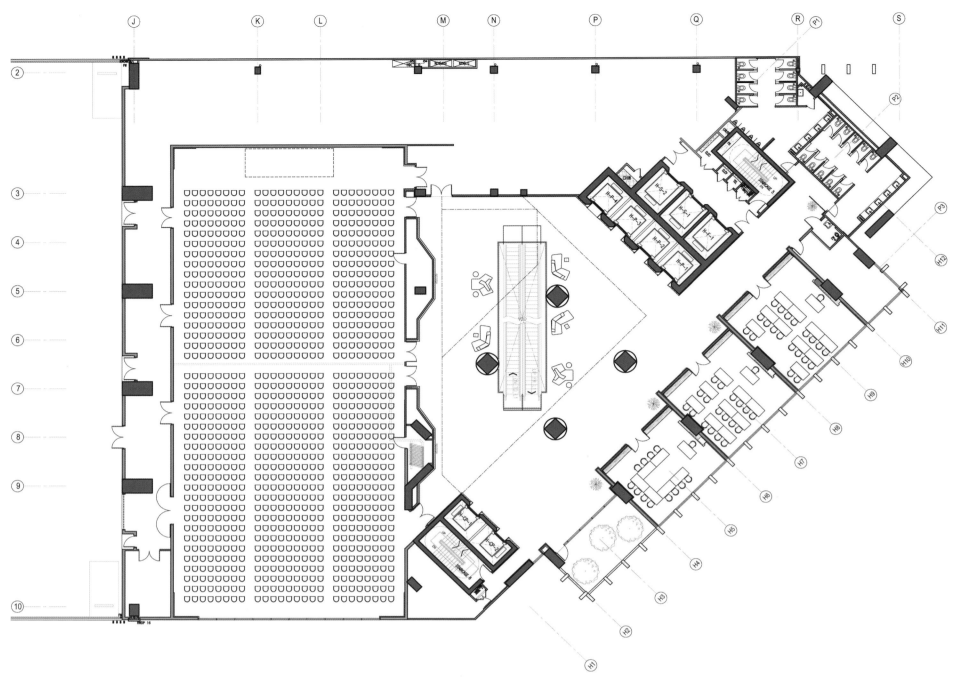

ALOFT KL

都市化设计混合地方特色是本案的设计风格。

这是目前世界上最大的一间雅乐轩酒店，大型宴会厅设计灵感来自于在公园漫步。超大的吊灯与地毯上木槿花叶的图案相得益彰。玻璃砖和墙面上回收利用的木饰板营造出现代的触感。

高挑空的房间设计灵感来自于工业阁楼。品牌标志色和充满生气的色调相混合，给空间带来活力。作为设计的重点，每个房间的床头板上都印制着艺术家 Antares 创作的卡通图画。这些插画以幽默的方式展示了马来西亚人的生活方式。天然的材料和纹理例如软木和木材，突出了艺术作品，创造出温暖的空间氛围。当地知名艺术家 Yusuf Gajah 创作的生动绘画作品，突出了戏剧性的灰色走道。

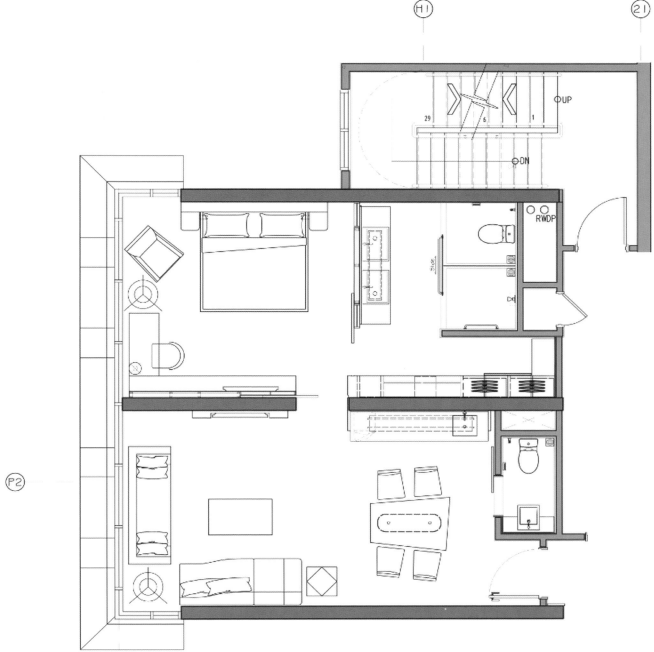

138　ALOFT KL

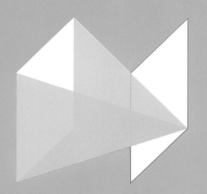

21st APIDA

LEISURE & ENTERTAINMENT SPACE
休闲/娱乐空间

LEISURE & ENTERTAINMENT SPACE 休闲 / 娱乐空间 **SILVER MEDAL** 银奖

Design Team / Kris Lin, Yang Jiayu
Location / China
Area / 4,000m²
Client / Times Property Co., Ltd.

TIMES BUND CLUB HOUSE

时代会所

Design Agency / Kris Lin Interior Design (KLID)

The meaning of life originates from the pursuit of the essence in life. Taking off the gorgeous cover and abandoning all the pomposities, it finally goes back to the value in people's heart and the truth of life. But art is unceasing exploration for human life, the repeated debating and rethinking, the practice and verification of life, which finally concludes as an idea and completes the art works. "The art of life, the life of art" is the core value of the design of the project which is aiming at to create a club of the sense of art.

Different from the 2D approach of drawing, the sculpture is more stereoscopic and the sculptors complete the works from three-dimensional angle. Designer of this project is trying to realize the design of the space through sculpture, which adopts 3D modeling instead of 2D model.

生活的意义源于对生命本质的追求。褪去华丽的外衣，摒弃一切的浮夸后，最终回归自我心中的价值，回归生活的真谛。而艺术是人类对生命不停的探索，反复的辩论与反思，是对生活的实践与印证，最终达成思想，完成艺术作品。"生活的艺术，艺术的生活"是本案设计的核心价值，旨在打造一个具有艺术气质的会所。

不同于绘画的平面式（2D）呈现，雕塑更加立体，雕塑家在3D空间中完成作品的创作。设计师试图通过雕塑的理念完成本案的设计，使用了3D建模而非2D模型。

一层配置平面图
First Floor Layout

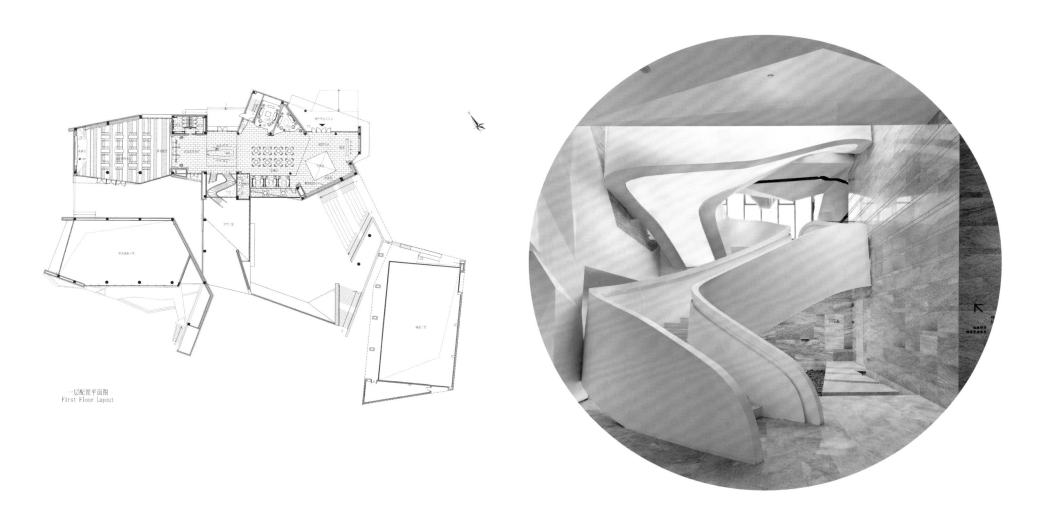

CLUB HOUSE 145

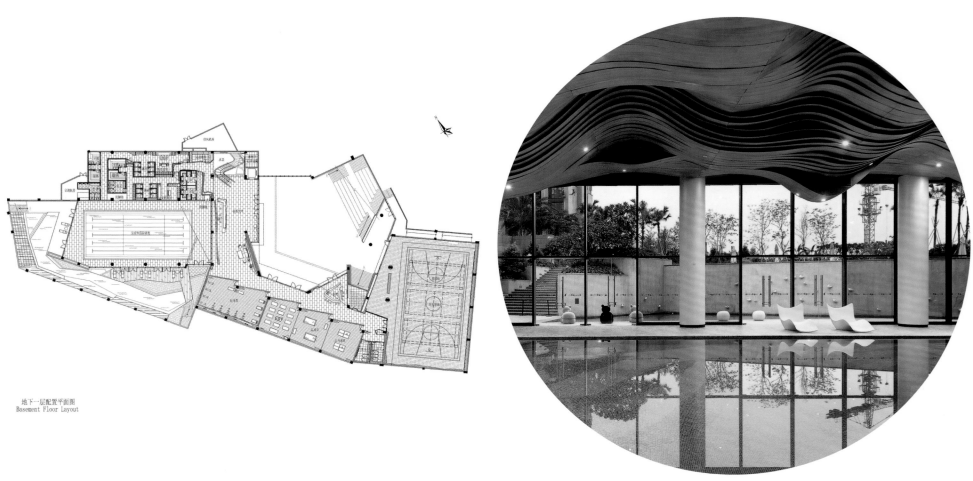

地下一层配置平面图
Basement Floor Layout

CLUB HOUSE

● LEISURE & ENTERTAINMENT SPACE　　BRONZE MEDAL
休闲/娱乐空间　　　　　　　　　　铜奖

DFS PSC LOUNGE

DFS 铂金服务俱乐部会所

Design Agency / pmdl Architecture + Design

The Platinum Services Club Lounge is located within the "Shoppes at Four Seasons" in Macau, amongst some of the most exclusive fashion retail outlets in the world. pmdl was commissioned to create an invitation-only space for some of DFS' very best customers, where they could relax and unwind during visits to Macau. The Lounge is approximately 800 m² and is organised as a series of lounge and entertainment spaces that open onto each other. Accordingly, the Lounge can be opened up to accommodate large functions or configured as separate intimate spaces, offering a variety of ways for guests to be pampered.

The spaces are grand in scale and finished with the highest quality materials such as timber paneling, marble flooring, and hand painted wall paper. The design embodies a true sense of place, drawing inspiration from the local heritage and culture of Macau. ●

　　铂金服务俱乐部会所坐落在澳门四季名店内，位于一些世界顶级时尚品牌直销店之间。pmdl 受 DFS 委托，为其贵宾客户创造一个在澳门期间可以休息和放松的空间，会所仅限于受邀请的人群进入。会所面积约为 800 平方米，以一系列彼此相通的休息室和娱乐空间的形式来布局。因此，会所既可以适用于一些大的活动，又可以配置成单独的私密空间，满足客人的各种需求。

　　空间壮观华丽，设计中使用了最好品质的材料如木嵌板、大理石地板和手工绘制的墙纸等。设计体现了真正意义上的地方意识，灵感来自于澳门当地的文化传统。

Design Team / Simon Fallon, Thom Silvius, Vania Contreras, Travis Backhouse
Location / Macau, China
Area / 800 m²
Client / DFS Cotai Limitada

PSC LOUNGE

149

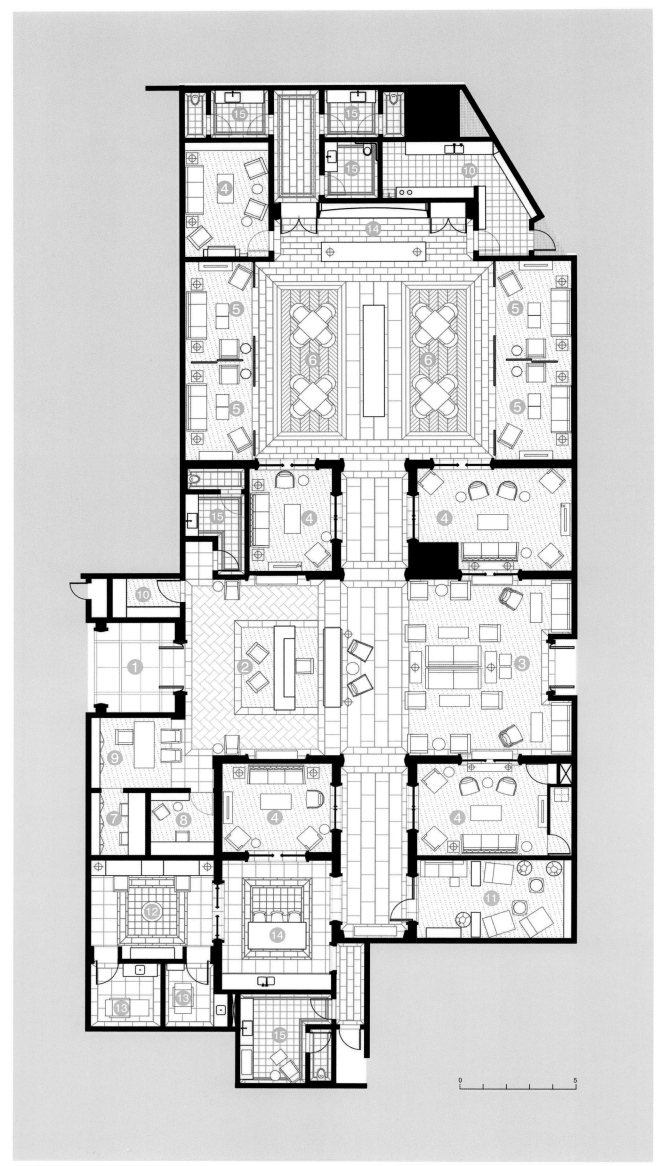

1	Entry
2	Reception
3	Grand Room
4	Private Room
5	Semi Private Room
6	Dining
7	Office
8	Guest Office
9	Concierge
10	Store
11	Kid's Room
12	Beauty Area
13	Facial Room
14	Pantry
15	Restroom

LEISURE & ENTERTAINMENT SPACE　**EXCELLENCE MEDAL**
休闲/娱乐空间　优秀奖

Design Team / Virginia Lung, Ajax Law
Location / China
Area / 4,500 m²
Client / Hubei Insun Cinema Film Co., Ltd. and Korea Lotte Cinema

TIANJIN INSUN LOTTE CINEMA

天津阳光乐园电影院

Design Agency / One Plus Partnership Ltd.

The cinema is a China and South Korea collaboration, which inspires the design concept of an "animal athlete meet" to resemble the friendship among two countries. Cinemagoers are greeted in the lobby with a backdrop of 3-dimensional lines, or "tracks", in startlingly colourful aquamarine, blue, lime green, yellow and purple. Ticket desks are fronted by large cut-outs of wild animals — from an ostrich, rhino and elephant to a bear, sheep and dinosaur. The flamboyant theme continues down the corridor leading to theatres, giving visitors the impression they are racing down the colourful "tracks" against more cut-out creatures running down the walls. Corridor seating is against the backs of foxes and baboons.

这间电影院乃中韩合作，为了彰显两国之间的友谊，设计师想到以"动物运动会"的形式来表达主题。客人甫踏进大堂，便会立即注意到背景的立体"赛道"线条，一系列悦目的色彩包括海蓝色、蓝色、翠绿色、黄色和紫色映入眼帘。售票处缀以大型卡通动物剪影造型，如驼鸟、犀牛、大象、熊、羊及恐龙等。这个耀眼夺目的主题延续到走廊并通往放映厅，让观众恍如走上色彩缤纷的"赛道"，同侧面更多的动物剪影比赛一般。走廊的座椅被设置在动物的脊背上。

LOTTE CINEMA

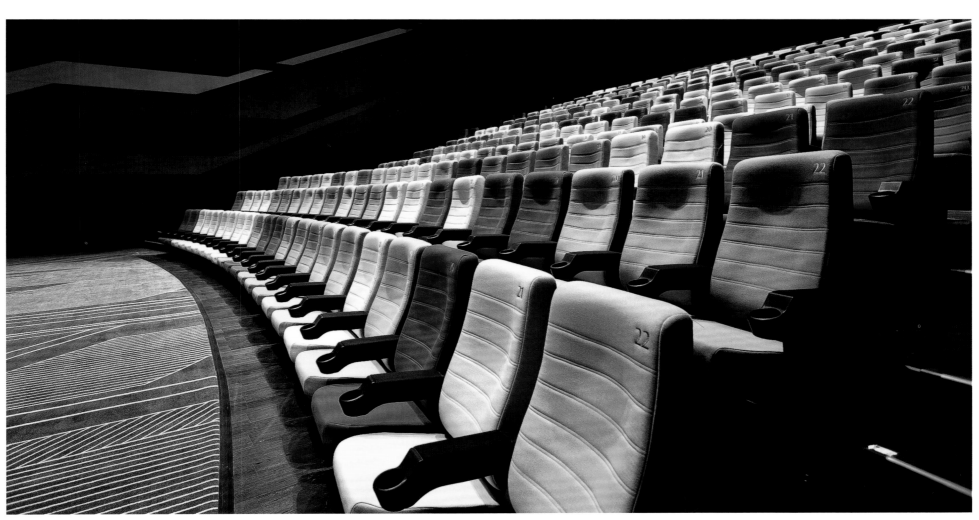

LOTTE CINEMA

LEISURE & ENTERTAINMENT SPACE 休闲／娱乐空间   **EXCELLENCE MEDAL** 优秀奖

Design Team / Zheng Chuanlu, Zhu Luxin
Location / China
Area / 550 m²
Client / Xiamen RongZeHe Investment Co., Ltd.

REME CLUB

REME 会所

Design Agency / Xiamen Share Opinion Design

REME is a membership-based club with the purpose of "music, international and communication". The owner requires a relatively low-key relaxing space. There are two zones in the space, the out zone and VIP zone. Designers divide multiple buffered transition space in order to achieve strict soundproof effect. Materials such as black walnut with saw notch surface, purple sheet copper and purple stone are used in the space. Custom-made water drop installation adds some intangible feeling into the space. ●

REME 是一间以"音乐、国际、交流"为宗旨的会员制会所，业主自身希望营造相对低调的释放空间。在空间的划分上主要分为外场及 VIP 内场两个区，隔音的要求相对严格，设计师为此分割了多个缓冲的过渡空间。在材质的运用上主要选用了黑胡桃木进行锯痕面的处理及做旧的紫铜板配上地面细心分割的山水紫石材。定制的水滴装置为空间增添了几分空灵。

● LEISURE & ENTERTAINMENT SPACE　　TOP TEN
休闲 / 娱乐空间　　　　　　　　　　十名入围

Design Team / Karr Yip Siu Ka, Wilson Lee Hau Pun
Location / Hong Kong, China
Area / 4,000 m²
Client / Youth Hostel Association

MOUNT DAVIS YOUTH HOSTEL

戴维斯山青年旅馆

Design Agency / ADO ltd.

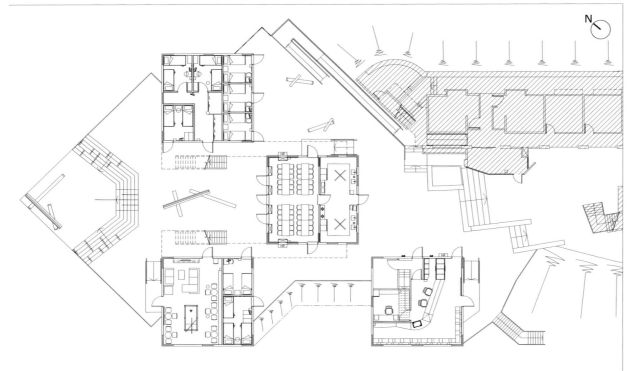

The hostel is located in the middle of a unique cultural landscape surrounded by ruins of batteries and barracks. All the new elements in the design from floor paving, furniture, lighting and signage capture the form of explosion: a symbol not just about destruction, but also about energy, movement and creation. This sense of dynamics matches perfectly the need to create a project for youngsters.

By turning it into an attraction point, the remoteness becomes a merit where you can sit in the comfortable common room, enjoying a 270 degree view of Victoria Harbor.

Being a renowned international hostel brand encouraging cultural exchange among youngsters worldwide, this project redefines the function of youth hostel by not just to provide accommodation, but also a living treasure box for them to explore the colonial history of Hong Kong. This opens up a new opportunity to combine hostel services and cultural tourism. ●

旅馆坐落在独特的文化风景区中，周围被遗迹废墟环绕着。设计中所用的所有新元素，包括地板材料、家具、照明和指引标志都捕捉了爆炸的形式；其不仅代表着毁灭，而且代表着能量、运动和创造。这种动态感与项目作为青年旅社的角色完美搭配。

旅馆所处的偏僻地点的不利条件成为一个吸引人的优点，旅客可以呆在舒适的普通房间中，享受270度的维多利亚港美景。

青年旅社作为国际知名的旅馆品牌，鼓励世界各地的年轻人之间的文化交流，这间青年旅社不仅为年轻人提供住宿，同时如同一个生活百宝箱，供他们去探索香港的殖民地历史。将旅馆服务和文化旅游结合到一起。

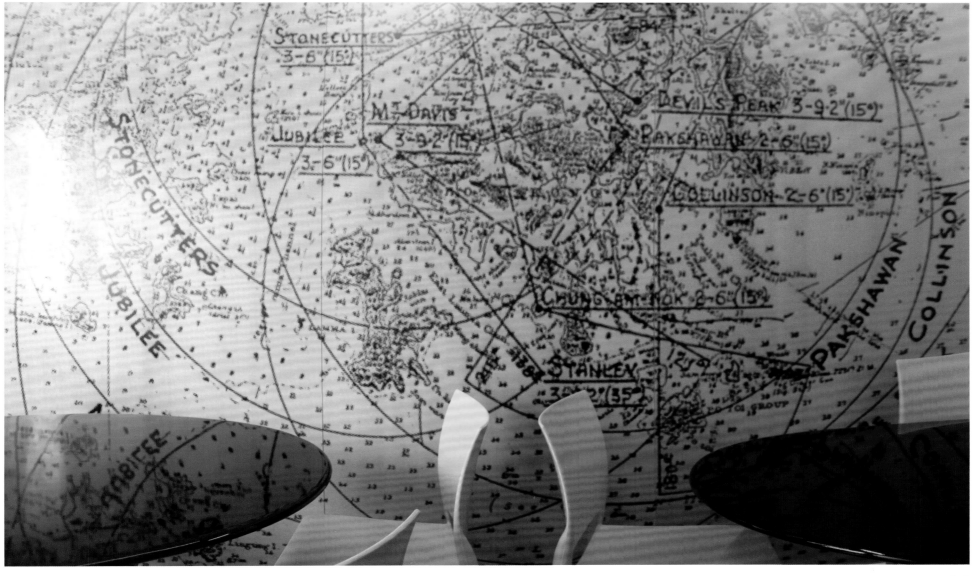

YOUTH HOSTEL

THE GYM

● LEISURE & ENTERTAINMENT SPACE TOP TEN
休闲 / 娱乐空间 十名入围

THE GYM BOX

健身拳击俱乐部

Design Agency / Creat8 Design and Management Sdn Bhd.

The Gym Box Malaysia is a health club that focuses on Mix Martial Arts, with a world class octagon boxing ring and a square boxing ring.

Concept of the design is Urban Industrial, to celebrate Mix martial arts in a different context and set up, to be different from conventional health club.

The best representation of industrial elements and recycled items include shipping containers, timber ply woods and timber crates, which can be seen throughout the whole space. ●

Design Team / Ninien Linggi (designer), Teo Chin How (project management)
Location / Malaysia
Area / 1,800 m²
Client / Sean Wong

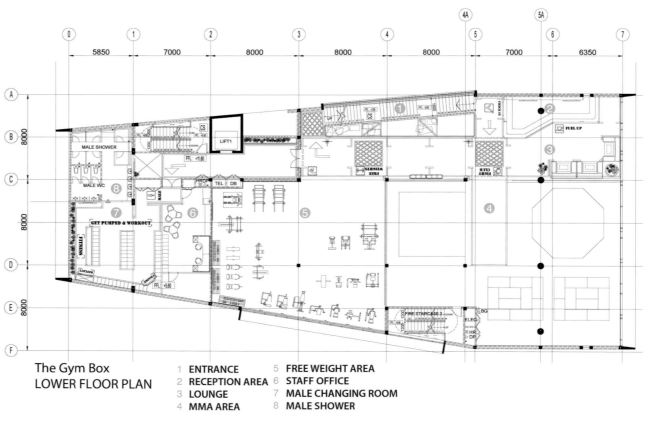

The Gym Box
LOWER FLOOR PLAN

1 ENTRANCE
2 RECEPTION AREA
3 LOUNGE
4 MMA AREA
5 FREE WEIGHT AREA
6 STAFF OFFICE
7 MALE CHANGING ROOM
8 MALE SHOWER

健身拳击是马来西亚的一间综合格斗健身俱乐部。内设一个世界级的八边形拳击场地和一个正方形拳击场地。

本案的设计理念是都市工业，赞美了综合格斗在不同环境中的运用及其创立，不同于传统的健身俱乐部。

工业元素和回收利用的物品例如集装箱、木板、木条箱等遍及整个室内空间，最好地诠释了设计主题。

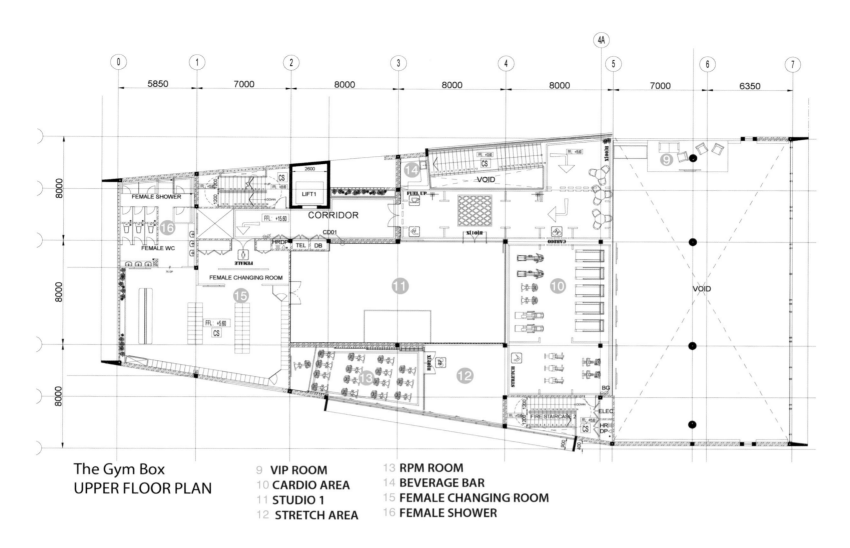

The Gym Box
UPPER FLOOR PLAN

9 VIP ROOM
10 CARDIO AREA
11 STUDIO 1
12 STRETCH AREA
13 RPM ROOM
14 BEVERAGE BAR
15 FEMALE CHANGING ROOM
16 FEMALE SHOWER

LEISURE & ENTERTAINMENT SPACE　　TOP TEN
休闲 / 娱乐空间　　　　　　　　　　　十名入围

CLUB AXIS THE WINGS

Axis 会所

Design Agency / Ronald Lu & Partners

The designers created an experience that consists of modern looks, with rich, sensual and tactile interiors. Club Axis is setting a new benchmark for other clubhouses alike how it presents users with an unmatched spatial experience that bridges architectural features with practical useable interiors. The sunken courtyard in the centre of the library demonstrates well how this piece of architectural feature can bring in ample natural light into the below ground spaces. The statement-like spiral staircase is also another successful attempt to express architectural interiors.

With its carefully considered planning, Club Axis is one fine example with pragmatic, experience-orientated architectural interiors for nowadays urban lifestyle. The spaces are neither far out, funky nor predictable nor over the top. The aesthetics is like decorative but not outrageous, tasteful with luxurious comfort in mind.

会所有着现代的外观和丰富奢华的室内设计，为其他类似的会所项目带来了全新的标准，为用户提供无与伦比的优质空间体验，将建筑特色与实用性完美结合。图书馆中心的下陷式庭院向人们展示了这种建筑特色如何为其下方空间带来丰富的自然采光。螺旋楼梯的设计是另一个成功的尝试。

会所经过精心的规划，建筑和室内设计注重实用性和用户体验，是现代都市生活方式的一个优秀范例。空间既不会过于激进、古怪，又不至于太过夸大。空间美学注重装饰又不过度，提供高雅而奢华的舒适享受。

Design Team / Alistair Leung, John Bachtiar, Sanford Ng, Jasmine Yau, Hely Chung
Location / Hong Kong, China
Area / 3,995 m²
Client / Sun Hung Kai Properties / MTR Corporation

THE WINGS

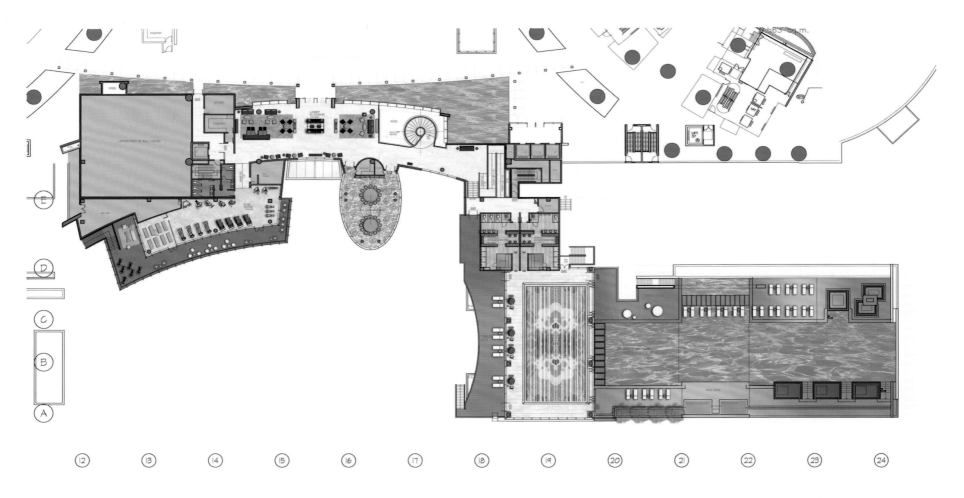

LEISURE & ENTERTAINMENT SPACE TOP TEN
休闲/娱乐空间　　　　　　　　　十名入围

AKOZO SALON & SPA

中国南京 AKOZO 资生堂沙龙会所

Design Agency / XJstudio

I. Entrance & Lobby area
1. Rosy exterior & open entrance hall
2. Curved marble walls
3. Rectangular marble concierge desk

II. Corridor area
1. Stylish walkway up to 50M
2. overlapping woods and mirrors
3. A scalp detection chamber combined with a show room
4. Merchandise is featured in boutique-style displays

III. SALON styling area
1. Long table vanity
2. Bookcase-style compartments
3. Separate shower rooms

IV. SPA conception area
1. Comfortable and spacious sofa reception area
2. Multifunctional bar counter

V. SPA area
1. More than 9 stylish themes

Location / China
Area / 695 m²
Client / MAGI Salon

一、入口大厅区

1. 绯红外观 & 开阔入口大厅
2. 大理石材弧形墙面
3. 长形大理石迎宾柜台

二、走道区

1. 长达 50m 时尚造型大道
2. 木质与明镜交迭
3. 结合商品展示的头皮检测室
4. 精品专柜式商品陈列

三、SALON 造型区

1. 长桌式镜台
2. 书柜式隔间
3. 独立包厢冲水区

四、SPA 接待区

1. 舒适宽敞的沙发接待区
2. 多功能水吧柜台

五、SPA 区

1. 多达九种风格的主题

● LEISURE & ENTERTAINMENT SPACE TOP TEN
休闲／娱乐空间　十名入围

ENZO CLUB

ENZO CLUB

Design Agency / INSPIRATION GROUP

This is a special club, a combination of function, decoration and science. Design concept of this space focus on fun, interaction and joy. Designer uses warm grey as main tone, and combines with modern design approach, creating an exclusive urban stylish lounge. Colors of purple, gold, blue and red enrich the layer of the space. With exalted taste and charming atmosphere, Enzo club becomes an ideal gathering place for Chinese and foreigner trendy. At the same time, it becomes a new entertainment landmark in Nanjing. ●

Location / China
Area / 1,073 m²
Client / Enzo Club

围绕"玩乐""互动""愉悦"三大设计需求，将功能与装饰、科技相结合，令 Enzo 独具特性。整间酒吧以暖灰色主调铺陈、简约时尚的设计手法，营造出专属都市时尚的 Lounge 格调，加入紫色、金黄、湖蓝、火红的光影，让空间层次更为立体与丰富。从视觉、嗅觉、感觉、听觉、触觉五感营建尊贵的品位与魅惑的气质，令 Enzo 成为中外潮流人士的聚集地，同时也成为南京城中娱乐新地标。

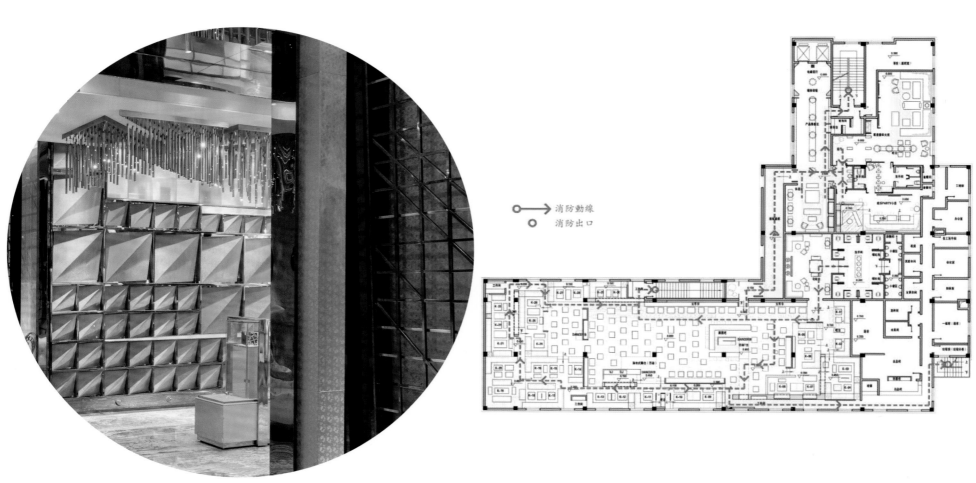

消防動線
消防出口

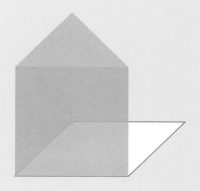

21st APIDA

LIVING SPACE
居住空间

● LIVING SPACE 居住空间   SILVER MEDAL 银奖

Design Team / Jacky Chang/ Jerry C. Dung/ Katie Yang/ Afie Huang
Location / Taiwan, China
Client / Mrs. Chen

RESIDENCE C

住宅 C

Design Agency / HHC DESIGN SOLUTION

To be one unit among a line up of townhouses, this project has its unique signature. To better fit the lifestyle of the family members, the interior was relieved from restraining internal borders while accepting liquid and layered layout provided by many design features like metal panels and glass.

If the displayed objects in this space are like the inlayed symbols, which are the hidden clues of the owner's memories, then accompanied by the expansion of the owner's living experiences, the symbols will act like the index puzzles to awake the memories. Therefore, through the combination and stacking of various elements, the designer forms a 3-dimensional frame in the open public space to create a "backdrop" for maximum possibilities of all the possible exhibits!

作为一列联排别墅的其中一间，本案有其独特的特征。为了更好地符合家庭成员的生活方式，室内设计摒除了压抑的室内界限，采用易变的分层布局，利用金属板和玻璃创造出许多设计亮点。

如果说空间内的摆件如同嵌入的符号，隐藏着居住者记忆的线索，那么随着居住者生活经历的延伸，这些符号就如同一个个拼图，可以拼凑唤起久远的记忆。因此，通过结合各种元素，设计师在公共空间中打造出一个 3D 的框架，创造出一个"背景"，用于尽可能多地去陈列这些"符号"。

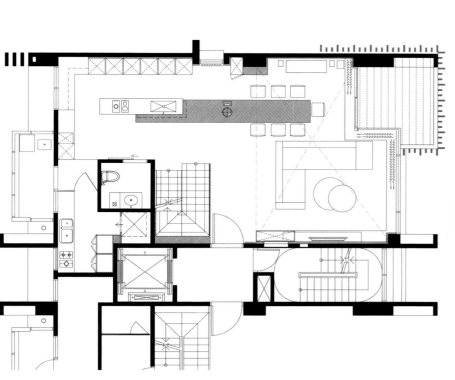

193

- **LIVING SPACE**
  居住空间

BRONZE MEDAL
铜奖

Location / Taiwan, China
Area / 251 m²

FLOWING INK

墨方

Design Agency / Shang Yih Interior Design Co.,LTD.

The design, with the stable and textured iron pieces, uses a stone with natural color and texture to surround into a rectangle in perfect ratio. The flowing lines of the marble on the fireplace divide the public space with some indirect partitions neatly; while the marble, named "ink landscape", which has the core values of the entire space, light and warm temperatures, is placed on the central axis of living room and study, living room and dining room. It brings out the warm color of raw wood calmly, and the hollow brick wall with the feel of earth.

以铁件的沉稳质感，将一块拥有大自然色泽与肌理的石材，围成一个完美比例的长方框型，壁炉上的大理石，流水般的线条将公共空间的利落刻划出间接性的区隔；而拥有整个空间的核心价值，与暖暖内含光的温度，名为"泼墨山水"的大理石，将之置于客厅与书房、客厅与餐厅的中轴线上，沉稳地带出原木温暖的色泽，与拥有大地触感的空心砖墙面。

● LIVING SPACE
居住空间

EXCELLENCE MEDAL
优秀奖

Design Team / Wanson Wan, Karen Mak, Isis Wong
Location / Hong Kong, China
Area / 250 m²
Client / Mrs Luk

LUK'S HOME

卢克的家

Design Agency / Wams Design Ltd

"Relaxation" is the one and only one request by the client.

Here is a Holiday house. Entrance is welcome door. A huge size door guides the guest from outside to inside.

Folding glass door are installed at the garden, blending the indoor and outdoor scenery while letting in natural sea breeze and sunlight, so that the apartment is filled with a very natural and relax feel.

The multi-colour sofa in the living room enriches the vitality and happiness.

"放松"是客户对于本案室内设计的唯一要求。

这是一间度假小屋。入口是迎宾门，巨大尺寸的门引导客人们进入空间中。

折叠式的玻璃门被安装在花园内，将室内外景观融合在一起，同时引入海风和日光，给公寓带来自然和放松的感觉。

起居室内多彩的沙发给空间增添活力和幸福感。

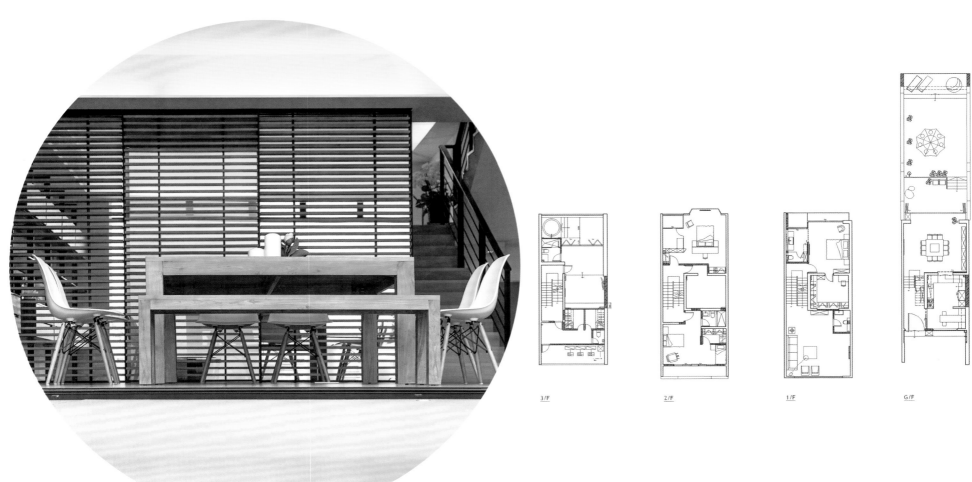

3/F   2/F   1/F   G/F

LIVING SPACE
居住空间

EXCELLENCE MEDAL
优秀奖

SKETCH PERSONALITY

速写延伸个性

Design Agency / W.C.H. INTERIOR DESING CO. LTD.

Between the structures of the top and bottom floors, one can capture a snapshot of work and leisure in a single moment. Broad and comforting visual angles are illustrated via perpendicular lines, and connect to form a cozy environment.

The design with neat and pure fundamentals entails the hobbies and personality of the inhabitants. Light softly shines into the room, casts long shadows through the open areas in the rear and front, and gives life to the open space.

Humanistic spirit gives a new life for this design through linear extensions that release an alternative form of modernism. Emotions penetrate and are combined with architectural concept to open a dialogue between space, environment, and its inhabitants.

从上下楼层的结构里，迅速捕捉工作和生活的意趣，透过垂直水平的线条，引申舒适开阔的视角关系，串连温暖舒适的情境氛围。

清浅基调，蕴含居住者的个性、喜好。前后开放的区域关系，让自然光线温柔轻缓地洒入，消长的光影，成为空间最佳的动态表情。

倾注人文精神，给予设计崭新生命，以延伸、通透的线面个性，演绎出另类的现代风格。作品渗透了情感并融入了建筑理念，建立了空间、环境与住户三者之间的对话。

Design Team / Wang Junhong, Chen Ruida, Cao Shiqing, Zhou Yijun, Huang Yunxiang, Lin Li
Location / Taiwan, China
Area / 72 m²
Client / Mr. Li

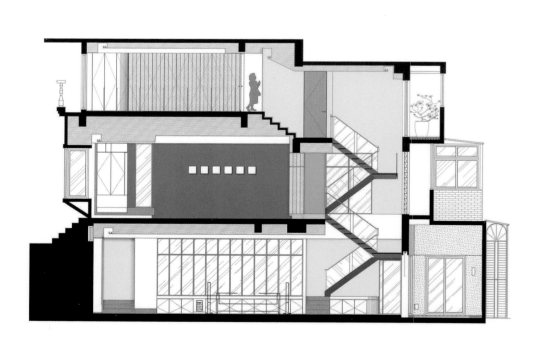

LIVING SPACE
居住空间

TOP TEN
十名入围

SPATIAL SEQUENCE

阅读空间的顺序

Design Agency / Design Apartment Co., Ltd.

This is a peaceful and poetic space. Space is where we live, while art is the cohesion of our life. Every artwork has its appropriate space. Between the dialogue of space and art, there are some hidden sequences to read the space. Materials enhance the depth of the space and reflect the inner peace, creating poetic atmosphere.

　　这是一个诗意的静谧空间。空间是容纳生活的器具，艺术是凝聚生活的感动。每件艺术作品都有其适切的空间居所。空间与艺术二者之间的对话互动，隐藏着一种阅读空间的顺序。透过强调原本材质的特色，加强空间的穿透层次，反射呈现出内在的宁静，让诗意的氛围于此呈现。

Design Team / Tang, Chung-han
Location / Taiwan, China
Area / 303 m²
Client / Mr. Liu

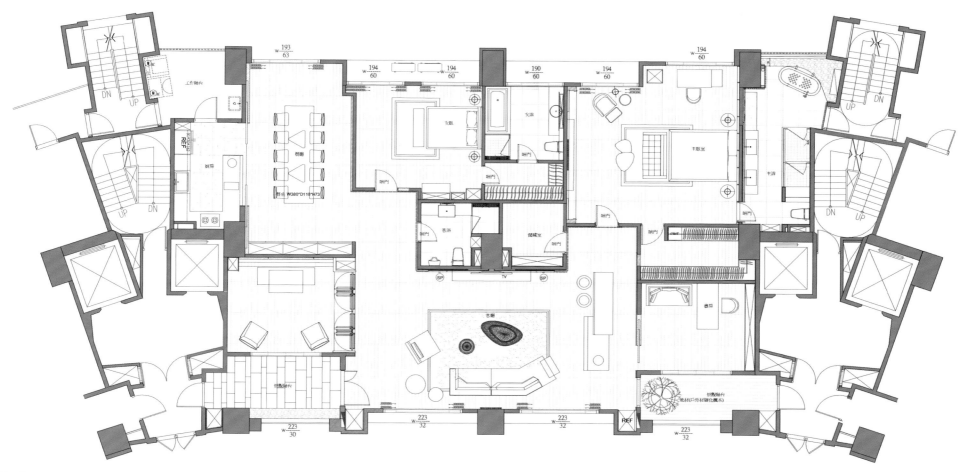

| LIVING SPACE | TOP TEN |
| 居住空间 | 十名入围 |

EXTEND DERIVATIVE

衍伸

Design Agency / Shang Yih Interior Design Co.,LTD.

"Inherited from the old building structure, the continuation of a love for Taiwan."

As the location moderator homeowners, the ideas put forward since the case straight to the point early design stage.

The designers deliberately retained the original construction of the structure of ceilings, brick walls, fireplaces, taking solid wood, makes up the strong taste of distressed styling, the set cement wipe out a gentle and calm gray tone floor, Irreplaceable sense of balance between the old and the new shape.

"承袭旧有建筑的结构体，延续一份对台湾的热爱。"

这是身为外景主持人的屋主，自本案初设计阶段就开门见山提出的想法。

设计师刻意保留原建筑之结构天花、砖墙、壁炉，用实木拼成仿旧味浓烈之造型；用水泥抹出一处温润沉稳的灰色调地坪，在新与旧之间，塑造出无可取代的平衡感。

Location / Taiwan, China
Area / 165 m²
Client / Ms. Xie

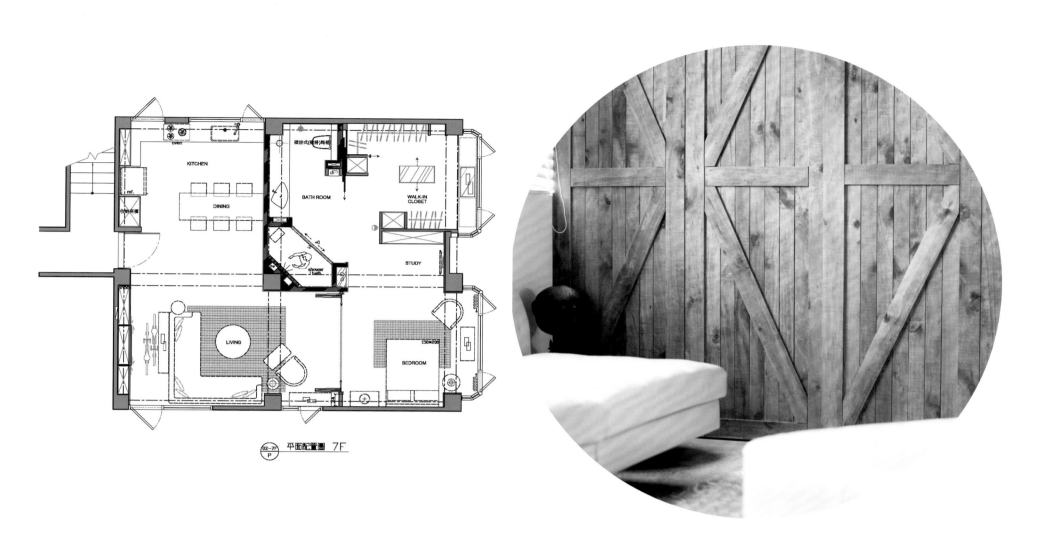

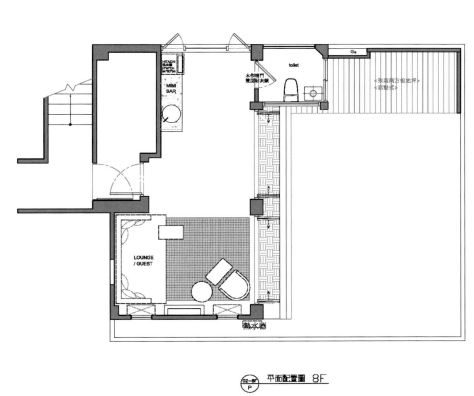

02-8F / 平面配置圖 8F

| LIVING SPACE | TOP TEN |
| 居住空间 | 十名入围 |

HOUSE IN SHATIN

沙田小屋

Design Agency / Millimeter Interior Design Limited

This design gave a boundless make over to the structure of an existing 40-year old house, transforming it into a comfortable and modern accommodation. The designer successfully divided this two-storey house into a garage, a living room, a dining room, a garden, two guest rooms, two guest bathrooms, one helper suite, a master bedroom suite with a spacious walk in closet and a study room.

设计师对一个有着40年历史的房子结构进行了整体大改造，将其转变成一个舒适而现代的居所。在两层的空间中，包含了车库、起居室、餐厅、花园、两间客房、两间客卫、一间帮佣套房、一间设有步入式衣橱的主卧套房，以及一间书房。

Location / Hong Kong, China
Area / 353 m²
Client / Cheung's Family

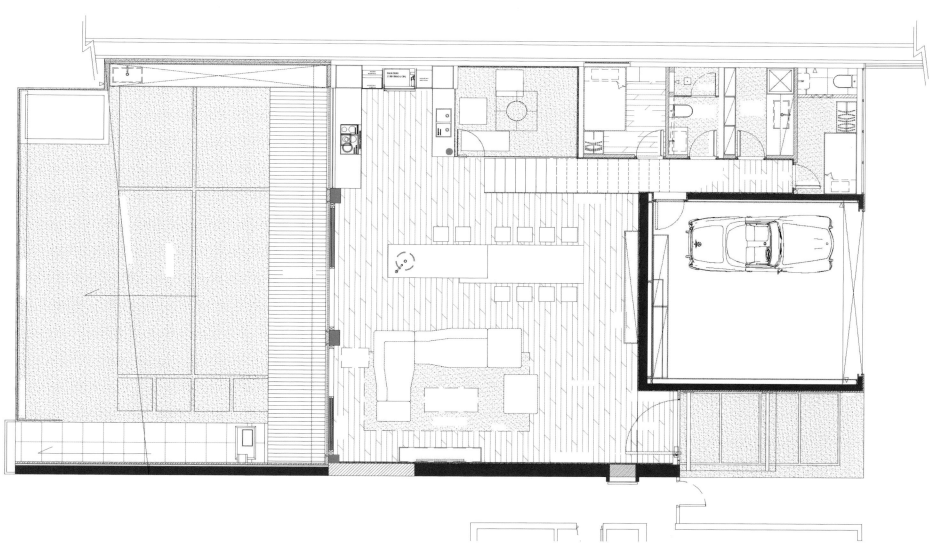

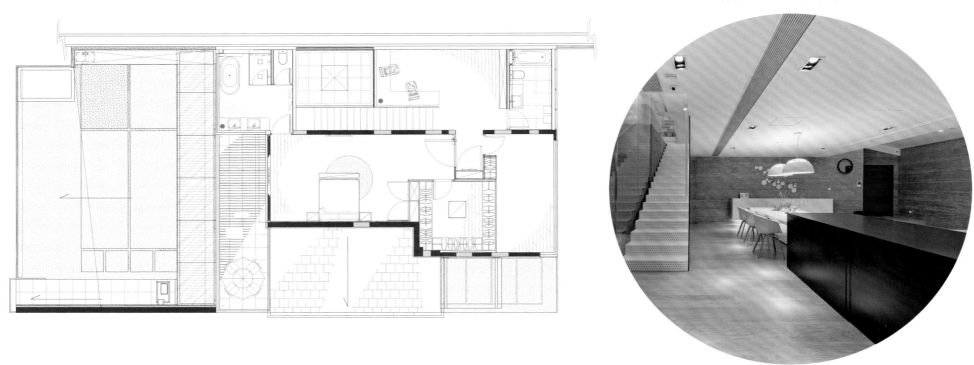

| LIVING SPACE | TOP TEN |
| 居住空间 | 十名入围 |

Design Team / Joey Ho, Kelvin Kong, Douglas Fung
Location / Singapore
Area / 538 m²
Client / Mr. & Mrs. Lim

BRACKET HOUSE

支架屋

Design Agency / Joey Ho Design Limited

Bracket House uses two simple curved planes: one vertical against a horizontal, to define its interior spaces. Designed from inside out, interior spaces are never enclosed and static.

The living space stretches beyond its internal boundaries, reaching to the far ends of the external boundary wall that frames the outdoor garden. Similarly, the dining space is semi-enveloped by the horizontal "bracket"; defined by a breathing green wall: a natural organic background. Nature is captured in print as a vertical wall covering along the staircase, connecting Bracket House from the basement all the way to the attic.

The soft curvature of the vertical "bracket" embraces the double volume attic space for the bedroom, just as the branches of a tree hover protectively over a tree house. The design's intention is to engage nature metaphorically and physically. It is the essential quality of Bracket House.

本案运用了两种简单的弯曲板，一个垂直一个水平，来定义其内部空间。设计由内至外，室内空间不存在闭合和静态。

起居空间延伸，越过其内边界，直至户外庭院的墙界。同样地，用餐空间被水平曲面半包覆着，会呼吸的绿色墙面成为天然的有机背景。自然的图景被用作垂直墙面的材料，沿着楼梯，将支架屋从地下室至顶楼联系起来。

垂直支架柔和的曲线，拥抱着双层楼高的顶部卧室空间，如同树枝盘绕保护着一个树屋。设计意图进行自然的比喻，这是支架屋的本质。

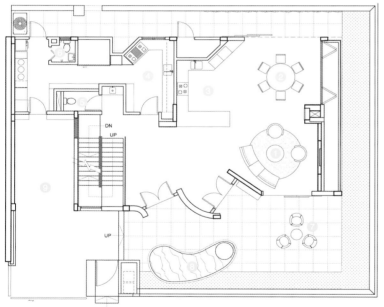

1. Living area
2. Dining area
3. Open kitchen
4. Kitchen
5. Powder room
6. Toilet
7. Outdoor living area
8. Pool
9. Car park

1/F floor plan

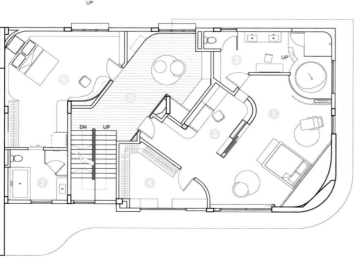

1. Master bedroom
2. Bathroom
3. Study area
4. Wardrobe
5. Gallery
6. Bedroom

2/F floor plan

- LIVING SPACE 居住空间 TOP TEN 十名入围

Design Team / Kenny Kinugasa-Tsui, Lorene Faure
Location / Hong Kong, China
Area / 117 m²

"BOATHOUSE" APARTMENT IN ABERDEEN

香港仔鸭脷洲"船屋"公寓

Design Agency / Bean Buro

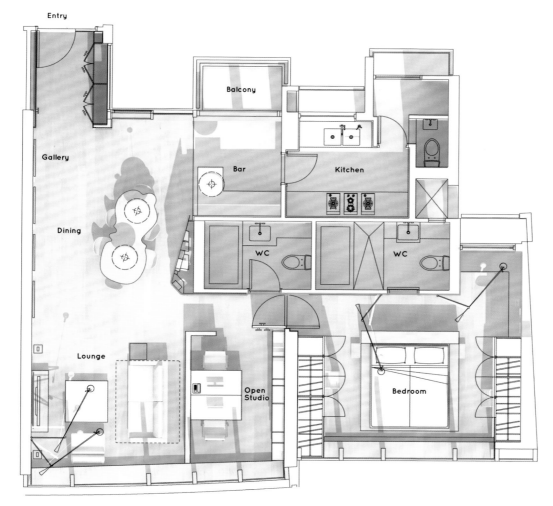

External view to Aberdeen

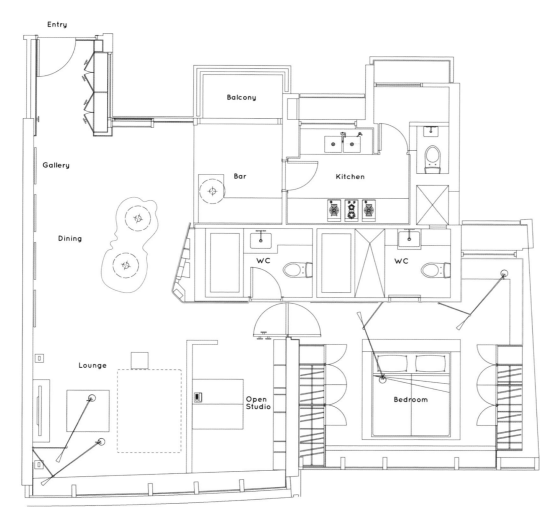

External view to Aberdeen

The three bedroom apartment in Aberdeen Larvotto was refurbished and transformed into a live/work/play apartment for a French couple with three cats.

The new concept demolished two existing partition walls to create a large bedroom and a lounge adjoining to an open studio space. The opened up spaces increases panoramic window views, thus allowing the external Aberdeen boating environment to be experienced inside.

Drawing inspirations from traditional French boathouses in Brittany, the main architectural concept was a continuous ribbon-like blue wall that "floats" and connects all the different areas of the apartment together.

The new timber flooring was conceptualised as a "cats landscape" that would rise and fall to provide different functions: it creates a beach threshold along the windows with an infinity-edge effect, an island dinning table to stage activities, a dynamic open studio, and a bed unit that faces the calm Aberdeen life.

位于香港仔鸭脷洲的一间拥有3个卧室的公寓被翻新改造，成为一对法国夫妇及他们饲养的3只猫生活工作玩耍的地方。

新的设计理念移除了原本的两堵隔断墙，创造出一个大大的卧室空间和一间毗邻开放式工作空间的休息室。开放的布局增加了全景窗的景致，让人在室内体验到外面的泛舟环境。

设计师从布列塔尼传统的法式船屋中得到灵感，本案的主体建筑理念是一个连续的缎带式蓝墙，"漂浮"着将公寓内各个区域连接在一起。

全新的木地板沿着窗户创造出一种无边界的效果。岛状的餐桌可用作进行各种活动。开放式的工作室充满活力，组合床面朝平静的香港仔，呈现一幅宁静的生活景象。

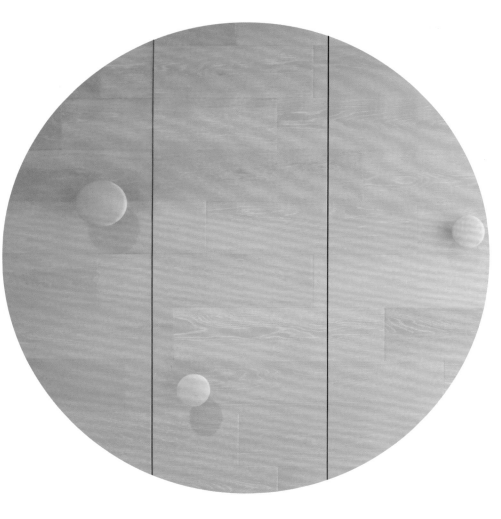

21st APIDA

SAMPLE SPACE
样板房空间

SAMPLE SPACE
样板房空间

SILVER MEDAL
银奖

海宁样板房

Design Agency / P1ang Studio Ltd

Design Team / Philip Tang, Brian Ip, Sim Pun
Location / China
Area / 430 m²
Client / Mingly Real Estate Corporation

In this 3-storey penthouse, designers use different kinds of leather, fabric and stone to create a soft and comfortable feeling over the flat. A few classic elements like picture frames and mirrors are incorporated with more contemporary touches.

In bedrooms, designers choose wallpapers and drapes that boast somewhat traditional motifs, such as a floral pattern, and combine them with furnishings and accessories that have distinctly sleek, streamlined silhouettes.

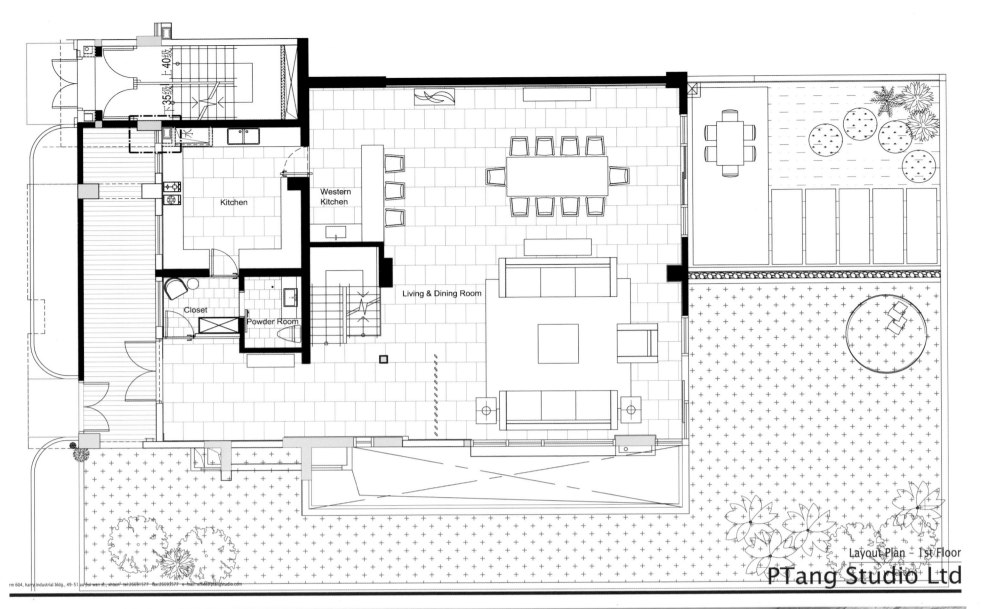

Layout Plan - 1st Floor
PTang Studio Ltd

在这个三层的空间中，设计师运用各种皮革、织物和石头创造出柔和舒适的整体氛围。诸如相框和镜子之类的经典元素与现代的氛围相得益彰。

卧室中，设计师选用了有着花卉图案的相对传统的墙纸和窗帘，与圆滑的流线型家具和配饰相搭配。

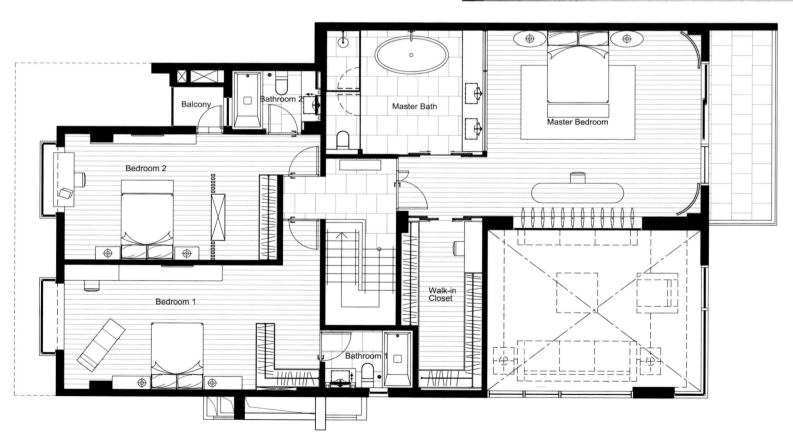

Layout Plan – 2nd Floor

Layout Plan - Basement Floor

SAMPLE SPACE 样板房空间	BRONZE MEDAL 铜奖		Design Team / YEN Bill CHEN-HSUN, Kellin Chen, Akon Li

Design Team / YEN Bill CHEN-HSUN, Kellin Chen, Akon Li
Location / China
Area / 64 m²
Client / Greenland Group

GREENLAND CHENGDU 468 MANSION SHOW FLAT

绿地成都东村 468 公馆样板房

Design Agency / Shanghai MRT Design Co., Ltd.

468 MANSION SHOW FLAT

When you enter the space, the first thing you will see is a streamline wall extending forward. White plasterboards are polished to adapt the entire architectural form through skillful technology, creating extended space. The main tone of the space is white, in order to create a visually bigger space within limited site. Three major materials that used in this project are white rock flour, timber and acrylic. Light wood veneer combines with white wall and fluid lights, creating cozy, clean and gentle living environment.

进入空间，首先映入眼帘的是一面延伸至前方的流体型墙壁。白色的石膏板塑造出延展状空间，通过构建工厂加工，工人娴熟的技术打磨成适应整个建筑形态的无棱角墙面。室内围绕着白色，力图在较小的环境中创造更大的空间。整体设计由三种主要材料构建——白色石粉、木料及透光亚克力。浅色的木饰面结合白色墙壁加上流线的灯光可以带来简约温馨、干净柔和的居家感觉。

SAMPLE SPACE 样板房空间   **EXCELLENCE MEDAL** 优秀奖

YINYI ISHAKES SERVICE APARTMENT

Yinyi Ishakes 酒店式公寓

Design Agency / Issi Design Ltd.

The interior design is based on a simple fashion style and emphasizes on the living usability. Direct and indirect lighting is largely used to offer a luxury but moderate private atmosphere instead of a complex and over-decorated one. The details including troffer design and convergent and smart use of the small space raise the exquisite and stylish value of the overall area.

本案室内设计的基调是简约时尚风格，强调了生活实用性。大量运用直接及间接照明创造出奢华而现代的私人氛围，而不会给人以繁复和过度装饰的感觉。设计细节包括天花板的暗灯槽设计，以及巧妙运用小空间来提升整体空间的精致时尚之感。

Designer / Andy Leung
Location / China
Area / 60 m²
Client / Shanghai Yinyi Developers Limited

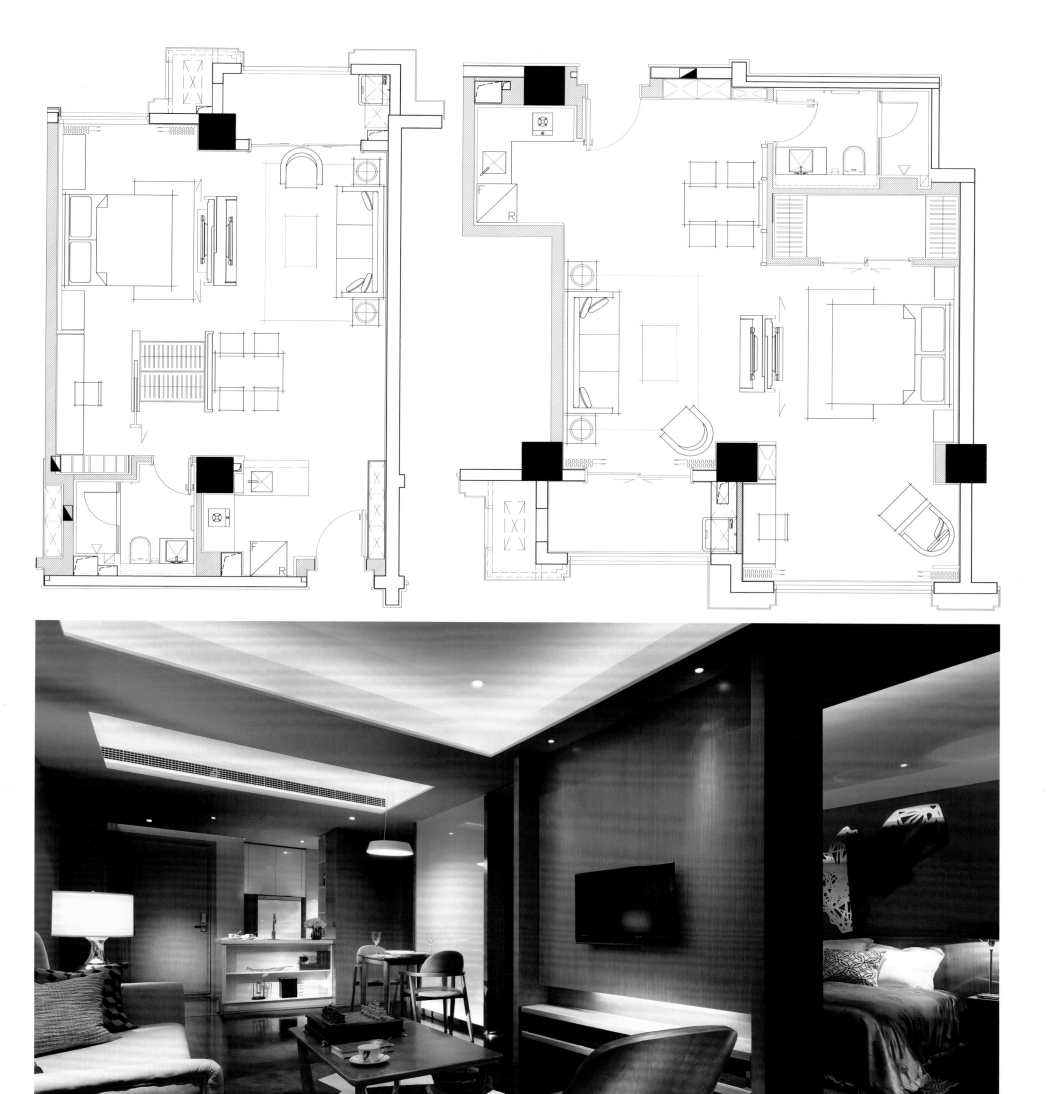

- SAMPLE SPACE 样板房空间　　EXCELLENCE MEDAL 优秀奖

Location / Taiwan, China
Area / 350 m²
Client / Mr. Chen

MOUNTAIN HOUSE

隐山居

Design Agency / Etai Space Design Office

Club is an outcome of material civilization and spiritual civilization development in modern times as well as the inevitable requirement of people along with the upturn spiritual and cultural living standards. Along with the increasing requirements of people about residential type, decoration and environmental, the concept of club becomes a kind of cultural supporting, blending with the development of humanistic quality.

会所，是现代物质文明和精神文明发展的产物，也是人们精神文化生活水平提高的必然要求。随着人们对住宅的类型、装饰、环境的要求的提高，"会所"概念成为一种文化的依托，也融合人文质素的发展。

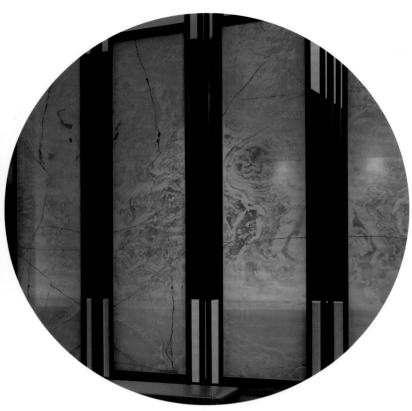

| SAMPLE SPACE 样板房空间 | TOP TEN 十名入围 | | | Designer / Han Song Location / China Area / 600 m² |

NINGBO DONGQIAN LAKE FONTAINEBLEAU YUE CLUB, PHASE 1

京投银泰宁波东钱湖悦府一期高端别墅——枫丹白露悦府会

Design Agency / Shenzhen Horizon Space Co., Ltd.

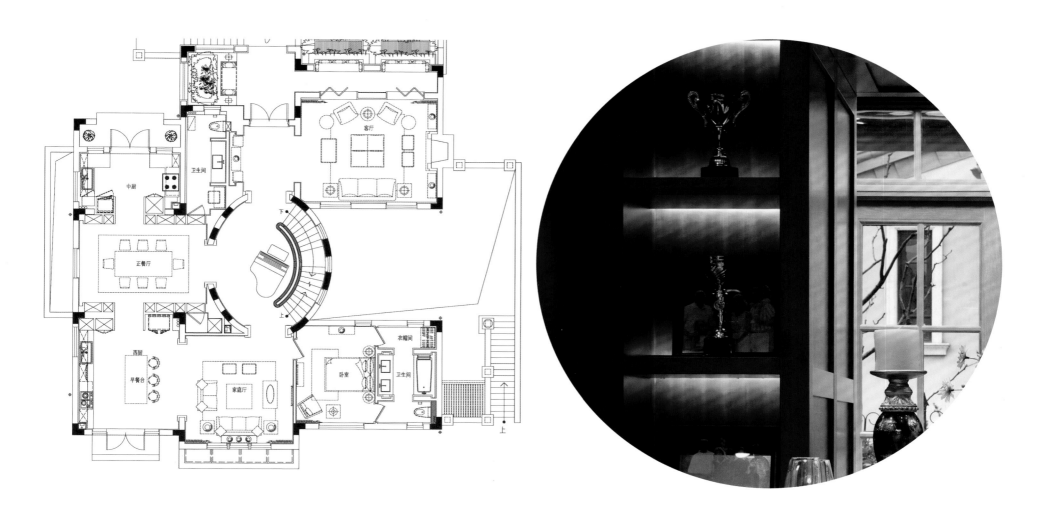

This project is an exclusive and luxurious private club, which provides high-end personal reception and entertainment services, from art salon exhibition to private party. The overall design of space integrates a great many British elements, such as the English wood panel, contemporary art pieces and the British band The Beatles' mementos. About furniture, designers chose the forefront of innovation British furniture brand HALO, which is synonymous with quality, vintage inspired leathers would bring you a fresh, exciting interpretation of classic.

As guests step into the club that their energy was awakens by. The lights are turned on to give the effect of growing daylight, and the curtain are slowly opened to extend the warmest welcome to distinguished guests.

Luxury Quality Smart Home System and other electrical devices are consistently equipped with top quality products. Both the Western and Chinese kitchens equipped with high-end facilities respectively. Special area for cigars and wine. All rooms are configured light scene control and music system, corridors intelligent light sensor system, automated hidden TV & sound system, washroom and cloakroom finger-print system. Lefroy Brooks sanitary wares bring fabulous classics experience of the last century.

Culture & Art Club comes with super butler service. Butler will show guests around the club and explain to them those details or behind stories of each piece of art, will take them to taste decent French wine, enjoy Havana cigar.

Two themes master rooms with completely different optical effects are separately under names of Fontaine and Bleau.

The club shared various services with Park Hyatt hotel, like sanitary service and chef service. In basement floor SPA area, may reserve Park Hyatt hotel's massager or Yoga instructor over and enjoy the exclusive moment.

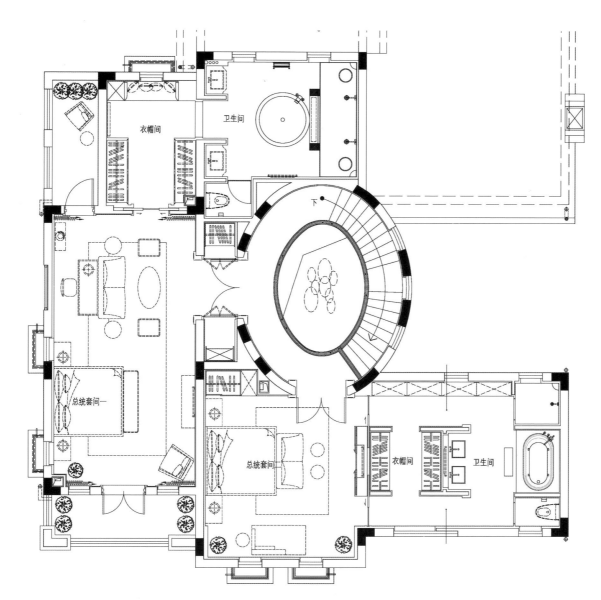

本项目为高端艺术沙龙性质的私人小型会所，提供高端的私人接待及娱乐服务、小型的艺术沙龙展览及高端私人Party。从室内设计到软硬件配置上都提供了顶级、高端的设施及环境。整体空间采用现代英式的木墙板装饰风格，配合英国HALO品牌的家私，以及当代艺术家的艺术作品和The Beatles的纪念物……于厚重、粗犷中带着一抹雅皮和时尚。

一、设计师希望客人进入会所的第一步就像生命被激活一样，灯光由暗变亮，窗帘徐徐开启，让客人感受到最热情的欢迎。

二、在硬件和智能化体系上追求英式的高品质传承。餐厨空间采用分离独立式中、西厨系统设计，各自配备不同的高端餐厨设备。各空间配备多种灯光场景模式和背景音乐系统，公共廊道感应式灯光控制系统，电视机隐藏控制系统，卫生间、衣帽间指纹电动感应开启系统。卫浴产品选用特别订制的英国老牌洁具Christo Lefroy Brooks品牌彰显贵族风范。

三、强调文化感和艺术气质。由真正的高级管家提供接待服务，为客户讲解设计的细节、每件艺术品背后的故事、每款家具的历史。客人还可以品尝地道的法国红酒，体验正宗的古巴雪茄。

四、就寝区设计了双主卧系统，并以枫丹、白露各自命题，展现完全不同的视觉效果和居住体验。

五、在软件服务上与柏悦酒店实现完美对接。如保洁服务、私人酒会、预约柏悦高级厨师到家中下厨，在地下室设计了专业的SPA空间，可预约柏悦SPA技师或瑜珈教练到家中提供专属服务。

SAMPLE SPACE 样板房空间　　TOP TEN 十名入围

VANKE SONGSHAN LAKE VILLA

万科松山湖中心别墅

Design Agency / Shenzhen Panshine Interior Design Co., Ltd.

This is a business villa with the area of 500 square meters, it can also act as a club. The target client is successful entrepreneur who is familiar with Chinese and Western culture. So designers adopt a design concept which is a combination of Chinese and Western elements. In terms of traditional elements, designers choose these elements by sensibility and handle them rationally before apply them to the space. On the other hand, modern pattern of manifestation creates living space fit for the lifestyle of modern people. People can earn edification in this place. They can leisurely experience the charm of tradition, and inadvertently feel the respective charm of Chinese and Western culture.

项目为兼具会馆功能的500平米商务别墅，目标客户群为自由游走于中西文化间的成功企业家，因此设计师选用中西合璧的设计方案。对于传统的元素，设计师经过了感性的提取，理性的加工与运用；另一方面，通过现代的表现形式营造出符合现代人生活的空间环境，让人们在内心得到熏陶后再从容感悟传统的魅力，境由心生，在不经意中感受中西文化各自独特的魅力。

Design Team / Zhou Jing, Zhou Weidong
Location / China
Area / 500 m²
Client / Vanke's Dongguan Branch

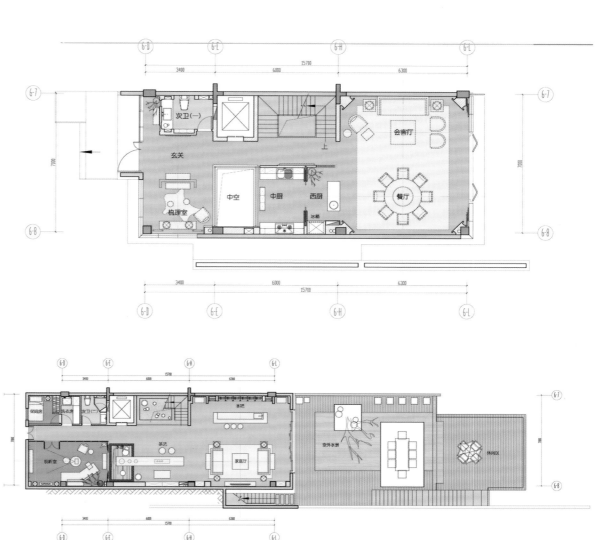

● SAMPLE SPACE
样板房空间

TOP TEN
十名入围

CHONGQING
VANKE
YUEWAN
SHOW FLAT A2

万科悦湾 A2 复式洋房

Design Agency / Shenzhen Matrix Interior Design

Location / China
Area / 405 m²
Client / Chongqing Vanke
Main Materials / Marble, Fabric Hard Package, Wallpaper

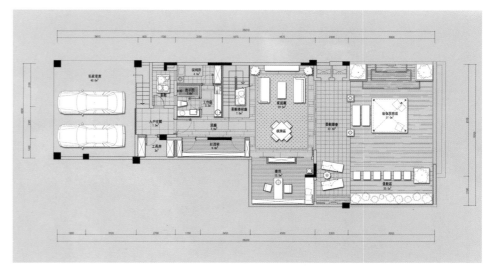

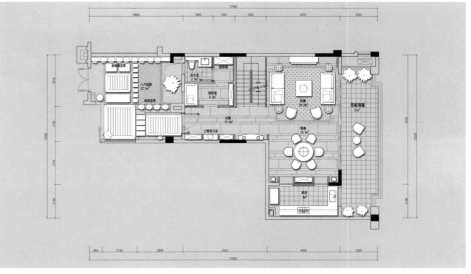

Black white and grey are the main tone in this space. Vivid lines and clear space layout create a set of Chinese ink painting which is elegant and steady. In terms of decoration details, it is natural and without too many colors. There are abundant Chinese classical elements in this space, such as traditional furniture, porcelain, antique aroma etc. All of these together, create a simple but elegant space with profound and lasting connotation of traditional culture, expressing people's pursuit towards quiet and steady lifestyle.

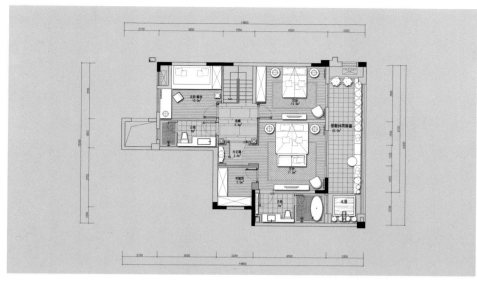

本案以黑、白、灰为主色调，以明快的线条和明朗的空间构图，手法娴熟独到，犹如一幅飘逸舒展的水墨写意画，清雅质朴，格调高雅，沉重内敛。在装饰细节上崇尚自然情趣，不用较多色彩装饰，融入了丰富的中国古典元素，传统家具、瓷器、古玩香薰等在不经意间打造了素雅古朴的至美空间，传达着传统文化的深远意蕴，表达人们对宁静致远的生活境界的追求。

SAMPLE SPACE	TOP TEN			Design Team / Philip Tang, Brian Ip, Harvey Tsang
样板房空间	十名入围			Location / Hong Kong, China
				Area / 550 m²
				Client / Yucca Investment Ltd

DE YUCCA HOUSE 15

雍雅山 15 号住宅

Design Agency / PTang Studio Ltd

High above the city of Shatin, the flat enjoy 180 degree of shimmering seascape. The sitting area connects to the private balcony and swimming pool.

Designers use different kinds of leather and fabric to create a soft and comfortable feeling. A few classic elements like picture frames and mirrors are incorporated with more contemporary touches.

In bedrooms, designer choose wallpaper and drapes that boast somewhat traditional motifs, such as a floral pattern, and combine them with furnishings and accessories that have distinctly sleek, streamlined silhouettes.

本案位于沙田，公寓坐拥 180 度绝美海景。休憩区与私人阳台以及泳池相连。

设计师运用各种皮革和织物创造出柔和舒适的整体氛围。诸如相框和镜子之类的经典元素与现代的氛围相得益彰。

卧室中，设计师选用了有着花卉图案的相对传统的墙纸和窗帘，与圆滑的流线型家具和配饰相搭配。

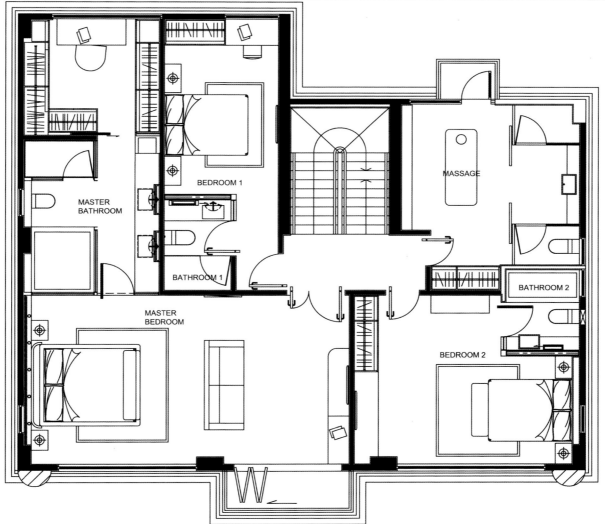

Layout Plan

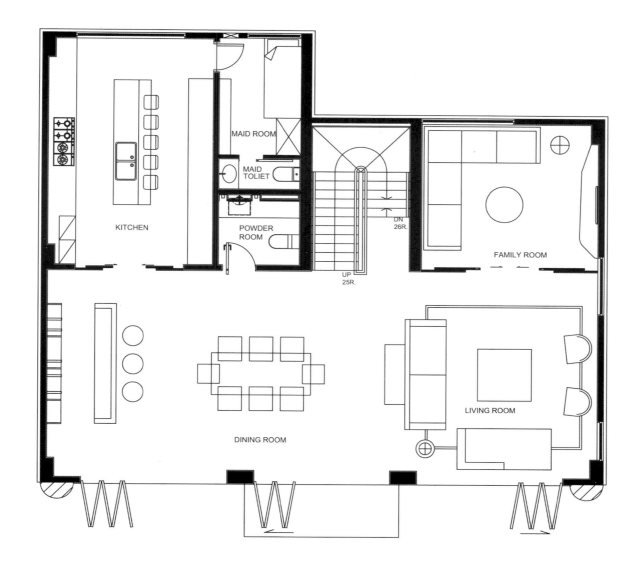

Layout Plan

267

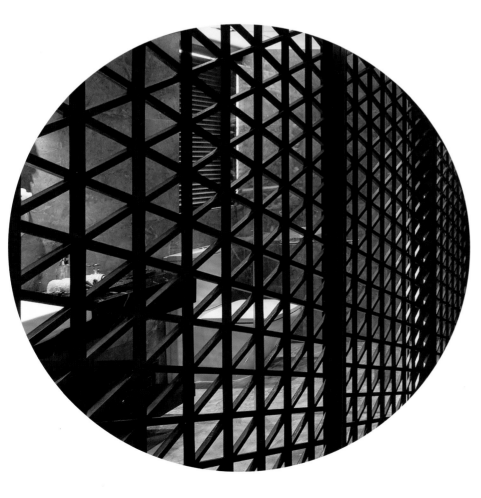

| SAMPLE SPACE | TOP TEN |
| 样板房空间 | 十名入围 |

Designer / Andy Leung
Location / China
Area / 63 m²
Client / China Resources Land (Shanghai) Limited.

CRLAND INTERNATIONAL COMMUNITY LOFT SHOWFLAT, NINGBO

宁波华润置地集团国际社区复式样板房

Design Agency / Issi Design Ltd.

The CRland international Community Loft Showflat has adopted the interior design positioning of an urban service hotel, a high-end elites' residence, so that it boasts a melody of moderate luxury by integrating the minimalistic style and the unique view and taste for the urban life from its target buyers. The limited space is divided into two floors, the first floor is dedicated to daily activities and dining, the second is designated for private bedrooms. The high-ceilinged living room, the TV cabinet and large bookcase on the ground floor and the French windows in all bedrooms, transparent master bathroom, all these are meant to show a small, exquisite and transparent space, clearly dividing public and private spaces. In terms of expression of materials, various advanced finishes (wooden cover, stone, mirror, wallpaper and hardcover, etc.) are against each other to demonstrate a moderate luxury taste of life.

华润置地集团国际社区阁楼样板房室内设计的定位是拥有都市酒店式服务的高端精英住宅公寓。空间的设计综合了极简主义风格和独特的视角，形成适度的奢华质感，满足目标客户的都市生活品味。有限的空间被划分成两层，首层被用作日常起居活动及用餐空间，二层则作为私人卧室空间。首层中有着高挑空的起居室、电视柜，以及大大的书架，所有的卧室都安置了落地窗，主卧配备了透明的浴室，所有这些结合到一起，组成了一个小巧、精致、通透的空间，公共空间和私人空间的界限分明。关于饰面材料，各种高级材料（木表层、石头、墙纸和硬包等）对比分明，展示了低调而奢华的生活品味。

COMMUNITY LOFT SHOWFLAT, NINGBO

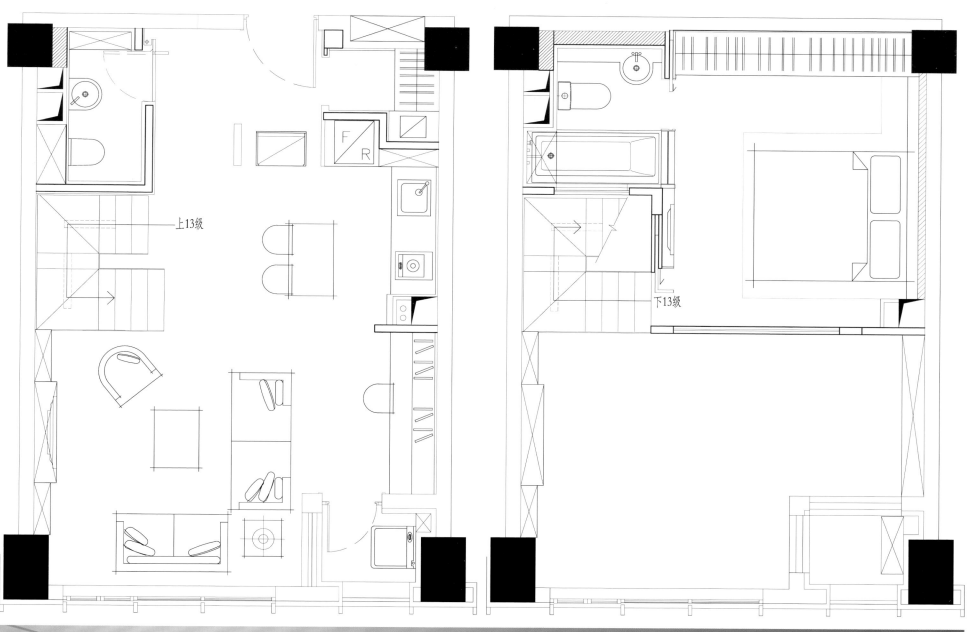

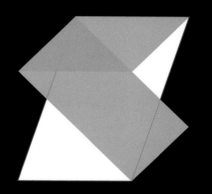

21st APIDA

INSTALLATION & EXHIBITION SPACE
设施 / 展览空间

● INSTALLATION & EXHIBITION SPACE   SILVER MEDAL
设施 / 展览空间   银奖

TIMES · BLOOM

时代 · 花生

Design Agency / Times Property Center — Design R&D Department

This space is originally a shop of two layers. After the integration and transformation it will be transformed into a vertical space. The three-dimensional composition is the matrix of this space. When the three-dimensional graphic design is being applied, it will combine the organics of fun and art, showing a visual sense of illusion and novelty. By resolutely discarding the former decorative materials, it will turn the space into a colorful carrier. The focus on the performance of color contrast will also produce rhytmical effects. From the space to the local, everything will bloom gloriously. ●

Designer / Gu Teng
Location / China
Area / 600 m²
Client / Times Property
Main Materials / Concrete, Aluminum Plate, Metallic Paint, Self-leveling, Glass, Metal

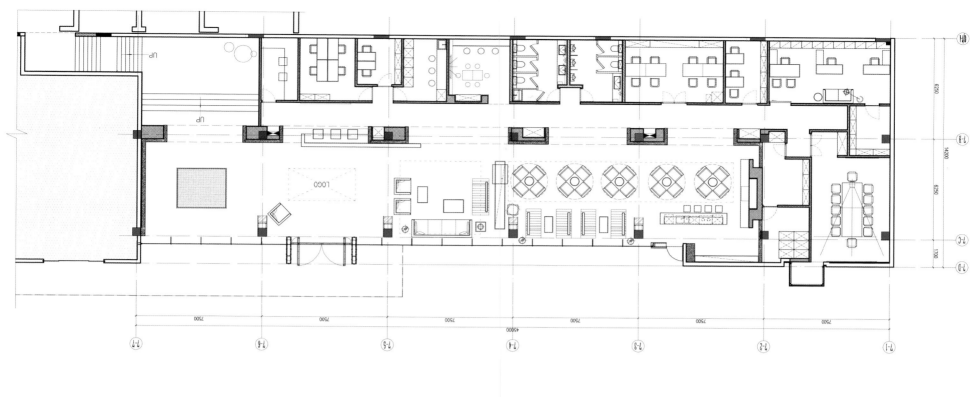

此空间本是两层的商铺，整合和改造后形成了一个纵向空间，立体的构成是此空间的设计母体，将平面设计运用到立体中去，让趣味与艺术有机地结合到一起，呈现视觉上的错幻感和新奇感，坚决地舍弃掉以往的装饰材料的运用，把空间作为色彩的载体，色彩的集中表现也在对比上产生韵律效果，从空间到局部每样东西都绽放着光彩。

● INSTALLATION & EXHIBITION SPACE　　BRONZE MEDAL
设施 / 展览空间　　铜奖

Design Team / Marc Brulhart, Seni Limpoon, Stephanie Chan, Ada Hung, Sukie Tsang, Lailing Lai
Location / Hong Kong, China
Area / 110 m²
Client / Swire Properties Limited

SWIRE PROPERTIES LOUNGE AT ART BASEL IN HONG KONG

太古地产贵宾厅——香港巴塞尔艺术展

Design Agency / Marc & Chantal

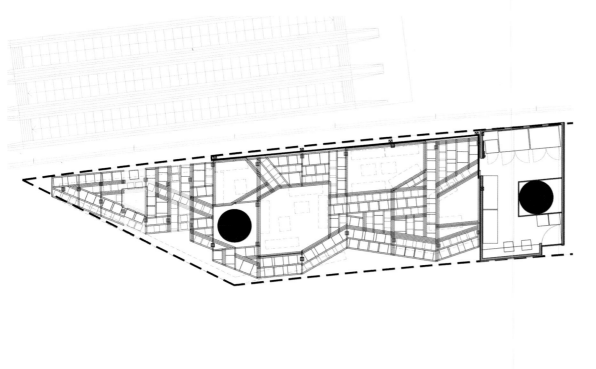

GENERAL LAYOUT - 1 : 75
Corporate Lounge ARTBASEL

IN HONG KONG

Hong Kong's famous density and intimidating crowded streets open themselves to a different perspective when traveling along its hidden corridors. Behind imposing walled entrances, the city's residents are connecting within neighborhoods through bridges, courtyards, tunnels and walkways that form rich networks in our daily lives. The installation at Art Basel invites visitors to discover the colours and textures found in communities across Hong Kong. The visual poetry found in the collage of materials, colours and shapes of these connections is seen in the 170 building models that feature photos of the architecture and people of Hong Kong.

The overhead timber framework symbolises the intertwined grids of the streets and was obtained from sustainable sources. When combined, the rectangular wooden supports were enlivened by the vivid photography to create a snapshot of this unique urban experience. ●

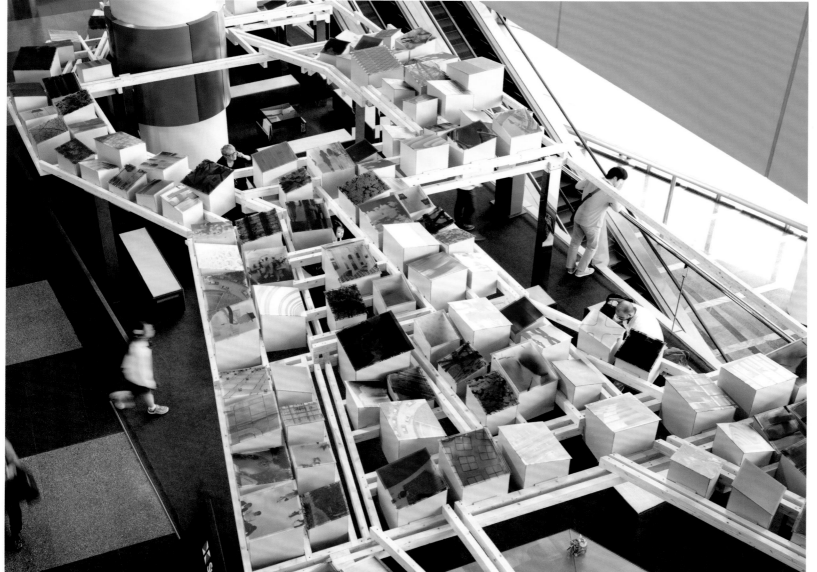

在香港最繁华最拥堵的街道，当你沿着其隐藏的走道行走时，将为你呈现出不一样的视角。在令人印象深刻的围墙之后，都市的居民们通过各种桥、天井、隧道和走道与邻里之间相互连通，它们组成了与人们日常生活息息相关的丰富网络。香港巴塞尔艺术展上的这个装置，邀请游客去发现和探索香港人生活的社区中的色彩和纹理。在这170栋建筑模型中，建筑和香港人的摄影作品，材料和色彩以及这些连接的形状，拼凑出一首视觉的诗歌。

木质的框架代表着错综复杂的街道网络。组合到一起后，生动的摄影作品给矩形的木质支撑带来生气，创造出这种独特的都市体验的快照。

- INSTALLATION & EXHIBITION SPACE EXCELLENCE MEDAL
  设施／展览空间 优秀奖

Design Team / Virginia Lung, Ajax Law
Location / China
Area / 1,215 m²
Client / Shanghai Forte Land Co., Ltd.

CHONGQING FLOWER & CITY SALES OFFICE

重庆花城售楼处

Design Agency / One Plus Partnership Ltd.

With the name "Flower and City", designers selected the "flower" and "kaleidoscope" as their main design elements. The designers simulate the effect of looking through a kaleidoscope by the use of irregular wall edges and different colors of ball decorations. Polygonal, angular, artificial, are how people would describe a kaleidoscope. Throughout the whole sales office, all the walls and ceilings are neatly finished with polygonal and triangular forms in white. Designers deliberately "shaped" the walls to uneven irregular triangles to depict abstract presentation of kaleidoscope. Spheres which are hanging on the wall are covered with seven different colors to symbolize the colourful flowers greeting kaleidoscope, scattered and disorderly, with the magic twist of the kaleidoscope. Designers use the spheres as an abstract method to illustrate how "flower" would look like under the influence of the kaleidoscope, highlighting the flower in its most basic structural elements.

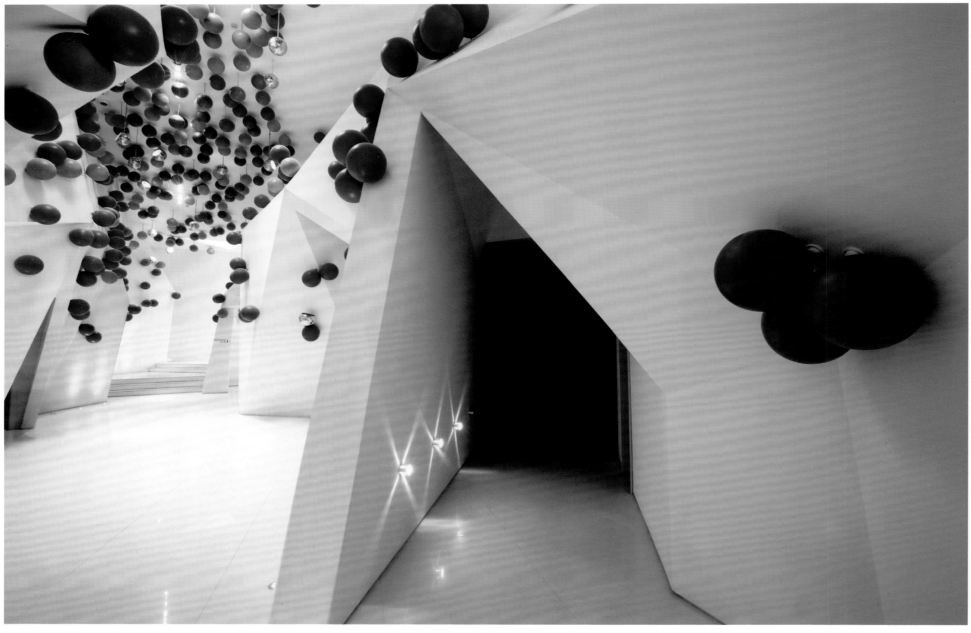

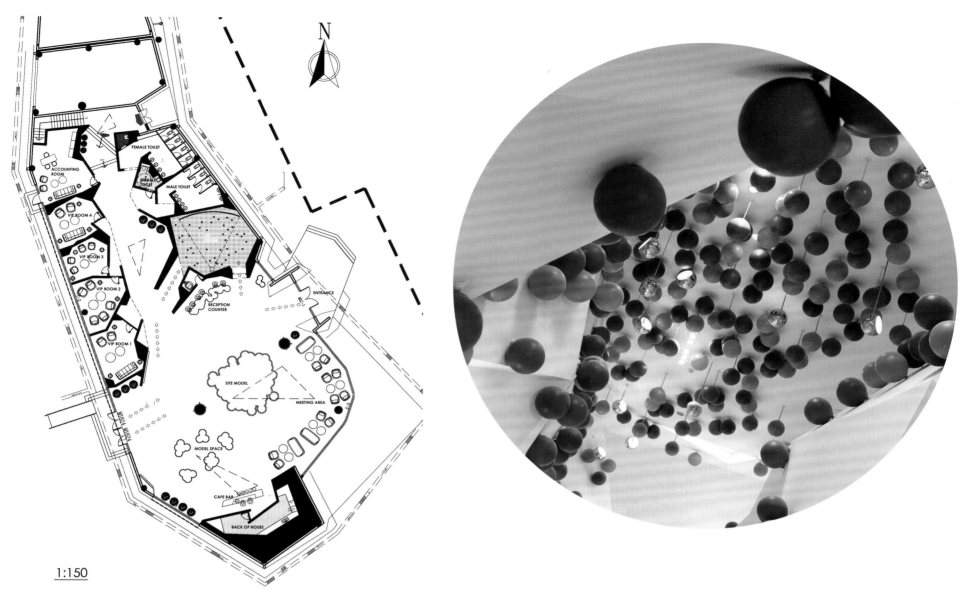

1:150

& CITY SALES OFFICE

项目以"花城"命名,选择"花"和"万花筒"作为设计的主要元素。设计师运用不规则的墙面切割和各种颜色的球体装饰来模拟人们透过万花筒所看到的景象。人们描述万花筒时所用的词汇通常是多边形、角度和人造。整个售楼处空间中,墙面和天花板都被设计成白色的多边形和三角形。设计师特意将墙面设计成不规则的凹凸三角形来描述万花筒千变万化的抽象形态。墙面上悬挂的七种不同颜色的球体象征着万紫千红的花朵映照在万花筒中,散乱而无序,正如扭转万花筒时所呈现的图像。设计师利用球体作为一种抽象的形式来阐述"花朵"在万花筒的影响下所呈现的景象,突出花本身最基本的结构要素。

● INSTALLATION & EXHIBITION SPACE　　EXCELLENCE MEDAL
设施 / 展览空间　　　　　　　　　　　　　优秀奖

Design Team / Peng Zheng, Shi Hongwei, Xie Zekun
Location / China
Area / 36 m²
Client / C&C Design Pavilion on Guangzhou Design Week

IDEA DOOR

多维门

Design Agency / C&C Design Co., Ltd.

C&C pavilion of Guangzhou design week is a multidimensional and synchronic space device. The windows and doors extended to four directions realize smart conversion and interaction inside and outside the display space, representing the enterprise concept of tolerance, openness and diversified development. By adopting the interactive display technology of augmented reality and the superposition of real environment and virtual environment, the enterprise design case inside the device brings about the conversion of display form from two-dimension to multidimension. ●

广州国际设计周"共生形态"馆是一个关于多维共时的空间装置，伸向四面的窗和门实现了展示空间内外的交互，同时也象征着包容、开放和多元化的企业理念。位于装置内的企业设计案例采用增强现实的交互展示技术，通过现实和虚拟环境的叠加，实现了展示形式由二维向多维的转换。

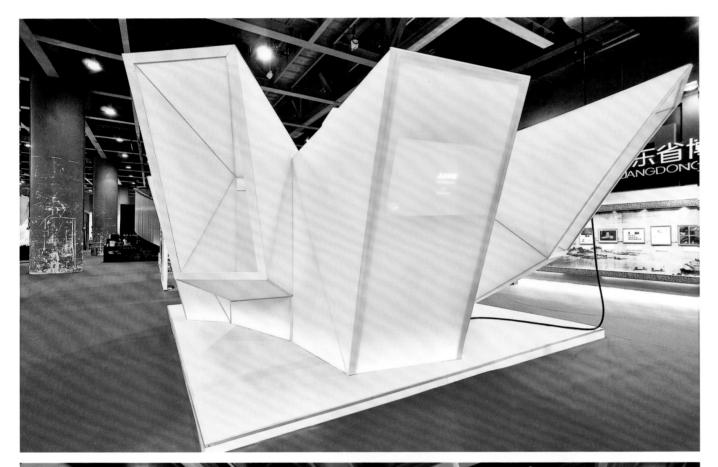

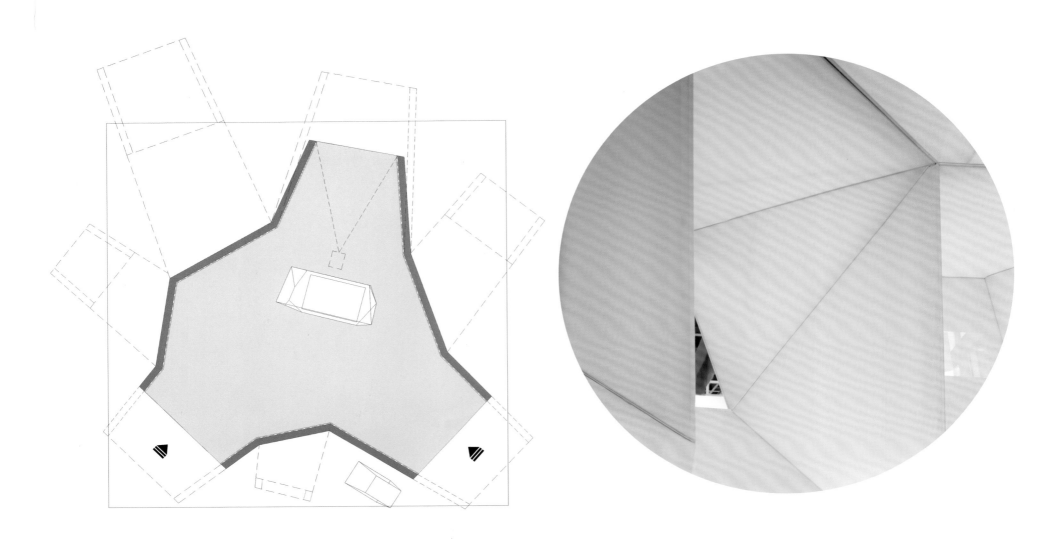

● INSTALLATION & EXHIBITION SPACE　　TOP TEN
设施／展览空间　　十名入围

Design Team / Lam Wai Ming, Alie Lam, Kent Wong, Zhang Xing, Tobey Ngo, Ferly Man
Location / Hong Kong, China
Area / Cultural Centre Piazza: 1,200 m²; JCCAC: 290 m²; Shatin Town Hall: 345 m²; Tsuen Wan Town Hall: 370 m²
Client / Hong Kong Photographic Culture Association

EYE TO EYE: JOCKEY CLUB SOCIAL DOCUMENTATION ROVING EXHIBITION

另眼·相看：赛马会社会纪实主题巡回展

Design Agency / Design Systems Ltd

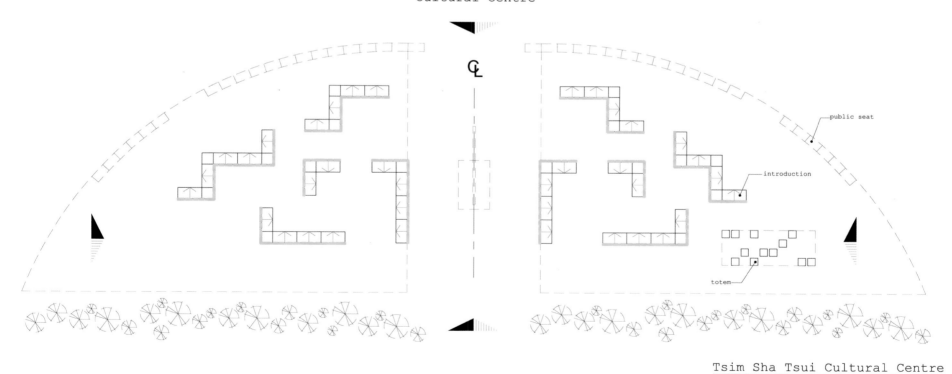

Tsim Sha Tsui Cultural Centre
Mirror side   1200 m²

This is the roving exhibition of the "Eye to Eye: Jockey Club Social Documentation Project", organized by Hong Kong Photographic Culture Association. The project offered photography training to novices from different walks of life, in order to encourage the public's social understanding by seeing what these participants see through their lenses. The venue is designed into a maze-like setting to let visitors meander and explore the different stories. The mirror surfaces of the display stands merge with the surrounding and make the display panels look as if floating in the space, thus putting the exhibits centre stage. It also creates a visual illusion where there is a mismatch between a real person and a reflected image of the visitors, and even with the protagonists in the photographs on display.

这是"另眼・相看：赛马会社会纪实摄影计划"的主题巡回展。该计划培训来自不同阶层和背景的摄影新手，透过他们的眼光去看他们认为最值得捕捉的画面。展场布局有如迷宫，让游人从中发掘每个故事。展柜侧的镜面设计令展板像悬浮于半空，既把注意力聚焦于展品，又营造视错效果，产生人的真身与镜像，乃至人与参展照片主角的错配。

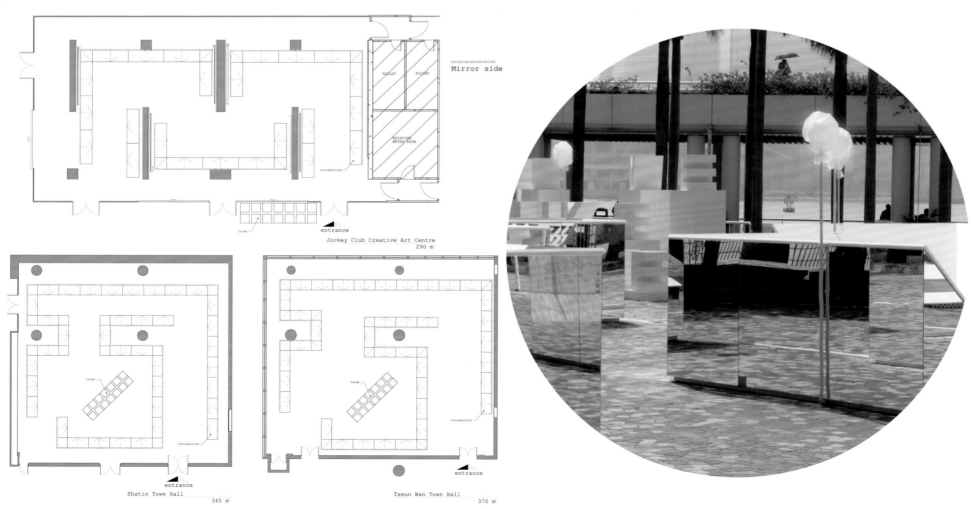

JOCKEY CLUB SOCIAL DOCUMENTATION ROVING EXHIBITION

● INSTALLATION & EXHIBITION SPACE　　TOP TEN
设施 / 展览空间　　　　　　　　　　　十名入围

NORTH GARDEN SALES PAVILION & GALLERY

Design Team / Liu Honglei
Location / China
Area / 1,400 m²
Client / Beijing Tianyang Real Estate Co., Ltd.

北京天洋北花园销售中心

Design Agency / BLVD

PAVILION & GALLERY

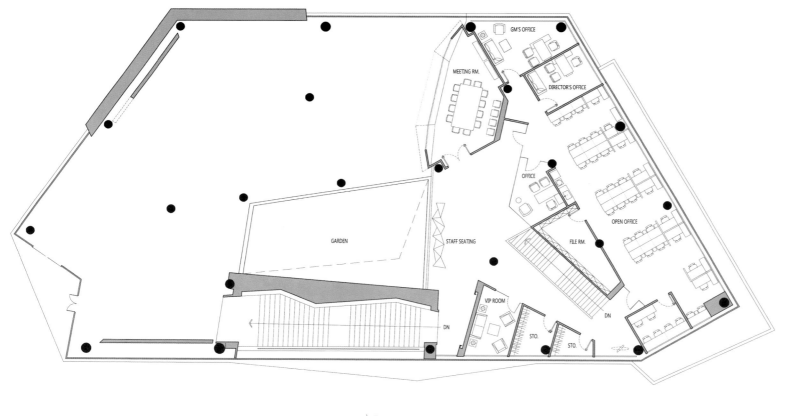

FLOOR PLAN-LEVEL 2

The design language, style and the ambience of the Sales Gallery and Showroom promoting contemporary residential properties to young, dynamic and upwardly mobile professionals in their twenties and thirties needed to communicate the trend and the lifestyle that was on offer. Prismatic geometry which is strong, bold, confident and definitive, much like the traits that are needed in today's professional world, reflecting the crystallization inherent in a rough diamond was the inspiration for this dynamic interior. Triangulated sculptured irregular surfaces - vertical, inclined or horizontal, fused together or free; whether used as a wall, window and ceiling or as a piece of furniture – defines, demarcates and dramatizes a strong visual identity and disseminates the initial conceptual idea. An abundance of natural light from all sides of the pavilion embraces the interior space creating further animated patterns across the surfaces as the sun moves across the sky. •

本案的设计语言和风格旨在为年轻人提供一种现代的居住选择，目标客户为20多岁至30多岁的正处于上升阶段的具有活力的人们，为其提供与时代潮流并行的一种生活方式。强烈、大胆、自信、明确的棱柱形几何形状，正如同当今职场中人们所需要的特质，也是这个充满活力的空间的设计灵感来源。三角形不规则表层，融合到一起抑或是单独存在，可以被用作墙面、窗户、天花板又或者是一件家具，呈现戏剧性的强烈视觉形象，体现了最初的设计理念。大量的自然光从展馆各个面涌入室内，伴随着太阳的移动，光线在表层上创造出更富生气的图案。

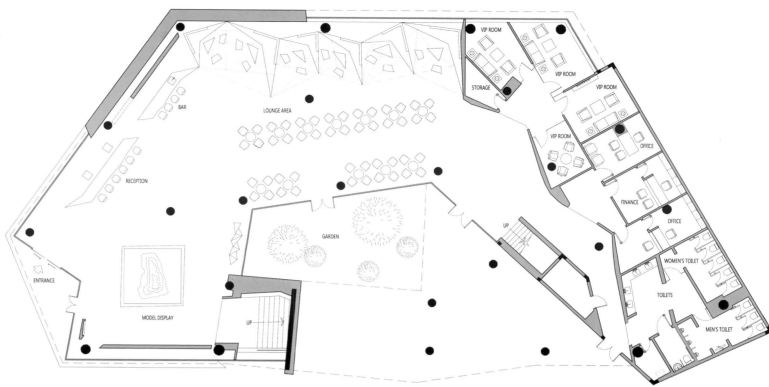

FLOOR PLAN-LEVEL 1

● INSTALLATION & EXHIBITION SPACE　　TOP TEN
设施 / 展览空间　　十名入围

Location / China
Area / 850 m²
Client / China SCE Property

DIALOGUE — ZHONGJUN SHANGCHENG SALES CENTER

《对话》——中骏商城销售中心

Design Agency / Xiamen East Towering Design

Strict geometric lines mark off the space relation effectively. Designer uses clean and direct approach to sum up complicated demand relation of the space. Use paper folding method to turn 2D lines into 3D space. The space structure with strong visual tension is very impressive. It combines with the sculpture of contemporary artist Gong Dong, performing a vivid drama in the space. It is a simple and perfect space.●

通过严谨的几何线条有效划分出空间关系。用简单直接的方式，高度概括复杂的需求关系。用当地民间艺术折纸的方法，把平面几何线条转化成三维立体的空间。使空间结构意气呼应，具视觉张力。结合当代艺术家龚栋的雕塑作品，仿佛上演着一场生动的话剧。作品寻求一种简单、直接、高度概括的极致之美。

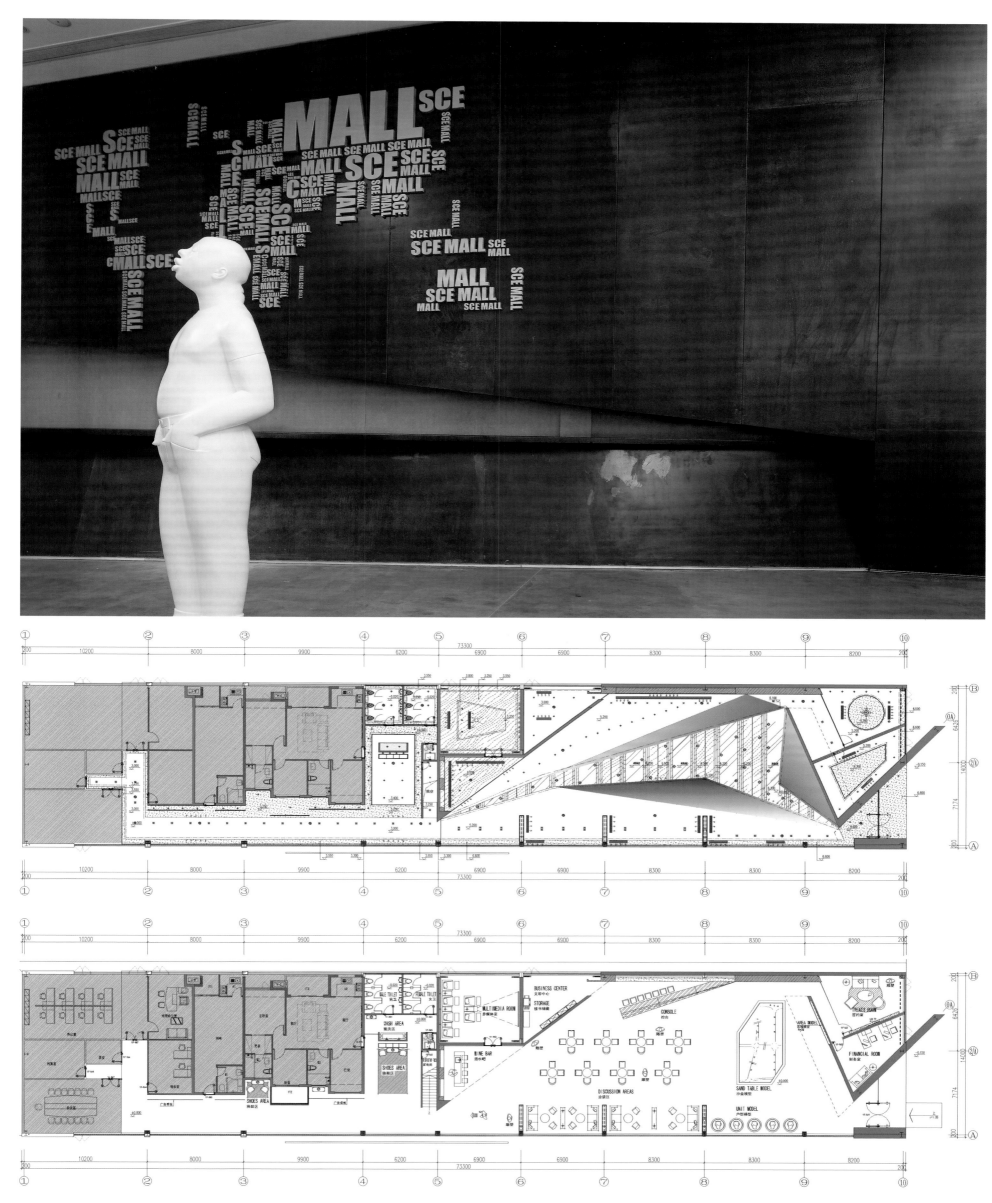

ZHONGJUN SHANGCHENG SALES CENTER

● INSTALLATION & EXHIBITION SPACE   TOP TEN
设施 / 展览空间   十名入围

Design Team / Kris Lin, Yang Jiayu
Location / China
Area / 1,200 m²
Client / China Vanke Co., Ltd.

HILLS SALES CENTRE

西山庭院售楼处

Design Agency / Kris Lin Interior Design (KLID)

It's a real estate sales centre. It is for sales of residential products. The project located in Taiyuan, Shanxi Province, China. Taiyuan is an ancient city with a history of 4700 years. It is in the middle of Shanxi Highland with many high mountains and hills. Hence the project is given a name Hills with the hope to reflect the human geography.

On 1st floor, it is mainly negotiating sofa area, bar, multimedia display area, model display area and reception area. It is hoped that the overall design also could embody the theme of Hills. Designers adopt sculpture and form the shape of mountain or palisades with three dimensional cutting. In model display area, the design concept is like a cave.

Go through the cave and the steps, it is 2nd floor. It is a tunnel, designers use three dimensional cutting to represent the cliffy rock in the valley. ●

本案是一个住宅楼盘项目的售楼处。项目位于中国山西省太原市，太原是一座具有4700年历史的古城，位于山西高原中部，有许多高山和丘陵，因此项目取名HILLS——西山庭院，也是希望能够体现这个城市的人文地理。

售楼处一层主要是洽谈区、水吧、多媒体演示区、模型展示区和接待区。为了在整体设计中突出山的主题，设计师利用雕塑，经过立体的切割，形成山形或岩壁的形状。模型展示区的设计理念是山洞。

穿过山洞，经楼梯上到二楼，进入"隧道"，设计师采用了立体切割来塑造出峡谷中陡峭的岩壁形态。

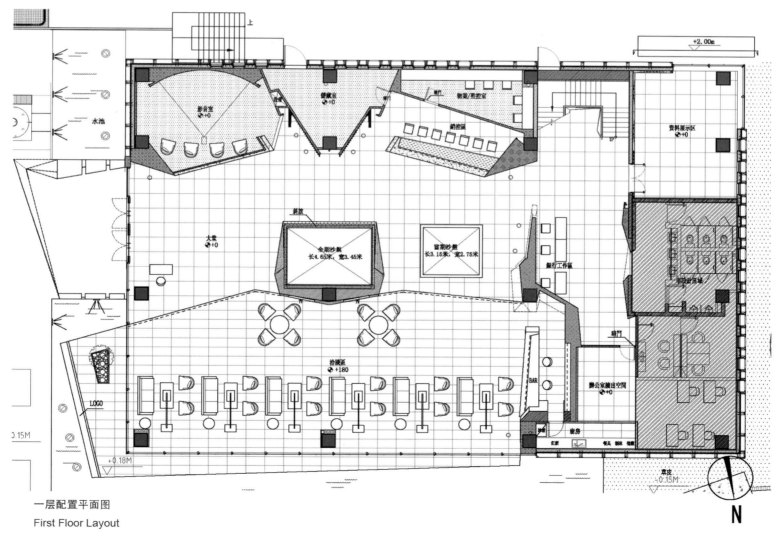

一层配置平面图
First Floor Layout

SALES CENTRE

二层配置平面图
Second Floor Layout

SALES CENTRE

● INSTALLATION & EXHIBITION SPACE　　TOP TEN
设施 / 展览空间　　十名入围

Location / China
Area / 850 m²
Client / Metro Land Corporation Ltd. Ningbo Qianhu International Center Development Co., Ltd.

YUE HOUSE

京投银泰宁波东钱湖悦府一期高端私人会所售楼处——悦府会

Design Agency / Shenzhen Horizon Space Design Co., Ltd.

The property is strategically located in Ningbo Dongqian Lake's natural scenic area, which is surrounded by Small Putuo and the Southern Song Dynasty Stone Inscriptions.

In terms of space: Both the internal and external space, in a traditional Chinese architectural sense, is enhanced by a sense of luxuriousness. In terms of visuals: The materials chosen were simple black and white color displays which were chosen to enhance beauty of Dongqian Lake. This was done to simulate the effect of watery ink on paper, and create beauty that could be compared to a water color painting.

The property provides a quiet, elegant, and easy living environment. Staying in this cozy cottage, and being surrounded by natural views, helps you to relax and unwind while escaping the hustle and bustle of city life. This place is ideal for people looking for enjoy nature, and live in a tranquil environment. ●

本项目依傍宁波东钱湖自然景区，独享小普陀、南宋石刻群等人文景观资源，地理位置无可比拟。

在空间上以中国建筑传统的空间序列强化东方式的礼仪感和尊贵感；在视觉上通过考究的材料和独具匠心的工艺细节，以简约的黑白搭配一气呵成，展现了东钱湖烟雨濛濛、水墨浸染的气韵。

本会所以浓浓的中式意蕴，展示一处幽静、高雅、洒脱的环境。身处这处清幽的别墅花园，可以隐逸其中，欣赏周围的开阔景色和新奇别致的幽景，静静地坐下来悠闲地聆听自然的声音，置身这种环境中，可以感到自己仿佛超脱凡尘，烦恼、杂念全部消失，非常舒心惬意。

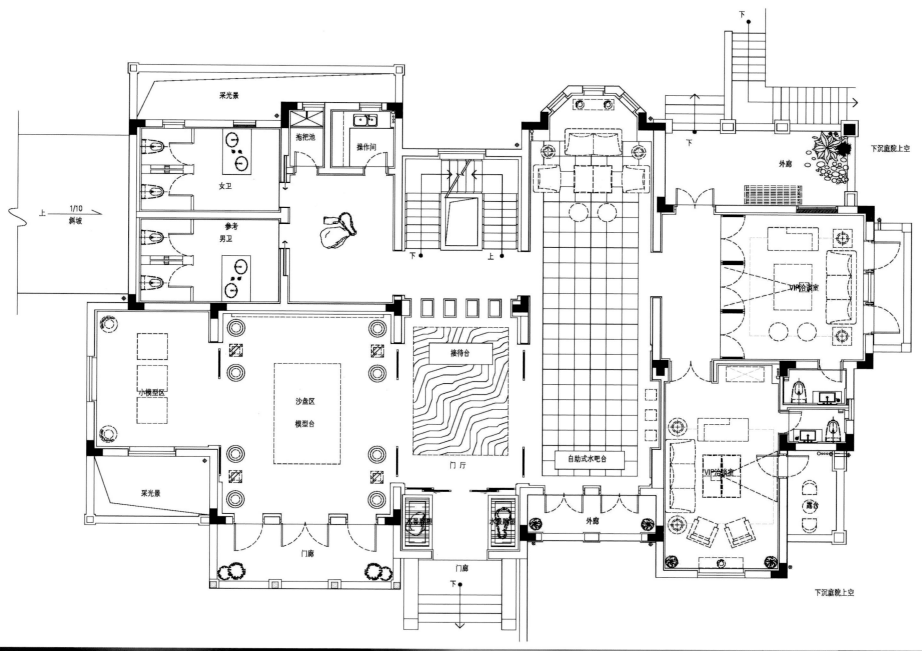

HOUSE 311

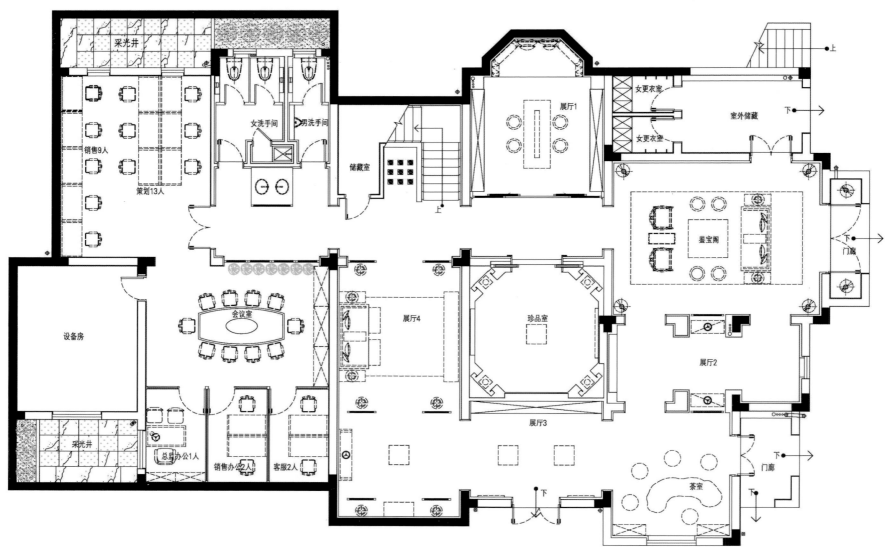

21st APIDA

PUBLIC SPACE
公共空间

● PUBLIC SPACE
公共空间

SILVER MEDAL
银奖

Location / Hong Kong, China
Area / 195 m²

CHAPEL

Chapel 教堂

Design Agency / Danny Cheng Interiors Ltd

Situated along the coastline, this white chapel impresses everybody with its marvelous and eye-catching appearance. The inspiration behind the design of the church ceiling comes from the honeycomb-like lines demonstrated by the AP700 model of the renowned car brand Lamborghini. The lining patterns of the ceiling match perfectly with the lighting effect, resulting in an amazing layered effect and make the design extraordinarily fashionable.

Chapels normally present a strong sense of purity and harmony. For this reason, the designer chooses pure white as its major tone, and makes appropriate use of simple lines and transparent glass in the design to highlight the tranquil and grand atmosphere of the chapel. The huge French Window not only serves to enhance the overall sense of transparency, but also draws in the scenery of sea and sky integration from the outside, bringing forth a romantic and relaxing sensation inside the building.

After the pool was filled with water, the chapel looks as if it is slowly emerging from the middle of the pool. The smooth water surface magically reflects the exact image of the building, displaying different visual effects with the changing scenery day and night. At the same time, with its coastal location and the refraction of lights, the chapel seems like a castle in the air in the distance at times.

On top of this, the designer skillfully hides the air-conditioning system and the light troughs of the building. Without anything redundant from the inside to the outside, the entire design was made simple and pure, creating a flawless and snowy chapel. ●

坐落于海岸线上的纯白色教堂，给人眼前一亮的感觉。天花设计的灵感源自名车兰博基尼 ap700 型号的蜂巢状线条，配合灯光效果，令整个天花设计充满层次感和现代感。

教堂给人一种纯洁和谐的感觉，因此设计师以纯白色作主调，再以简单的线条和通透清明的玻璃来配合教堂的宁静和庄严的气氛。大大的落地玻璃除了增加通透感外亦把海天相连的景致引进教堂内，令室内充满浪漫醉人的感觉。

当水池注水后，教堂像缓缓地浮现于水池的中央，平静如镜的水面把整个教堂清晰地映衬出来，除了日与夜的变化，还带来不一样的视觉效果。因临海关系，在光线的折射下，远眺时，仿佛"海市蜃楼"。

设计师亦细心地将冷气系统及灯槽位隐藏起来，整个设计从内到外没有多余的东西，干净利落，缔造出简洁无瑕的纯白教堂。

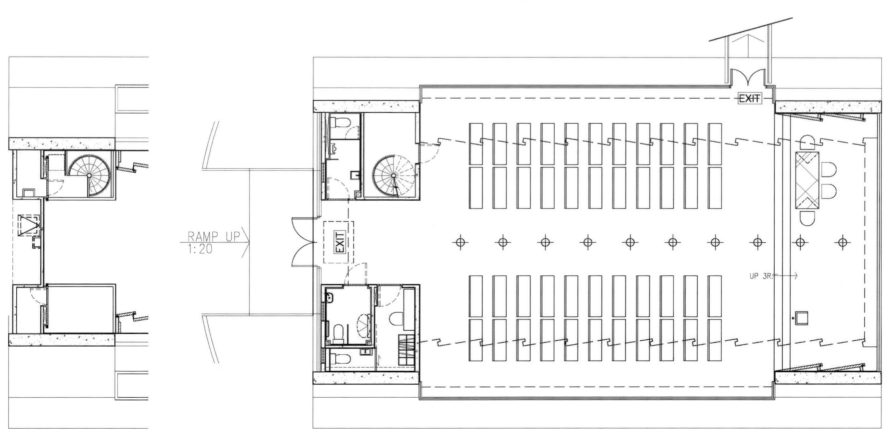

UPPER FLOOR   GROUND FLOOR PLAN

317

CHAPEL

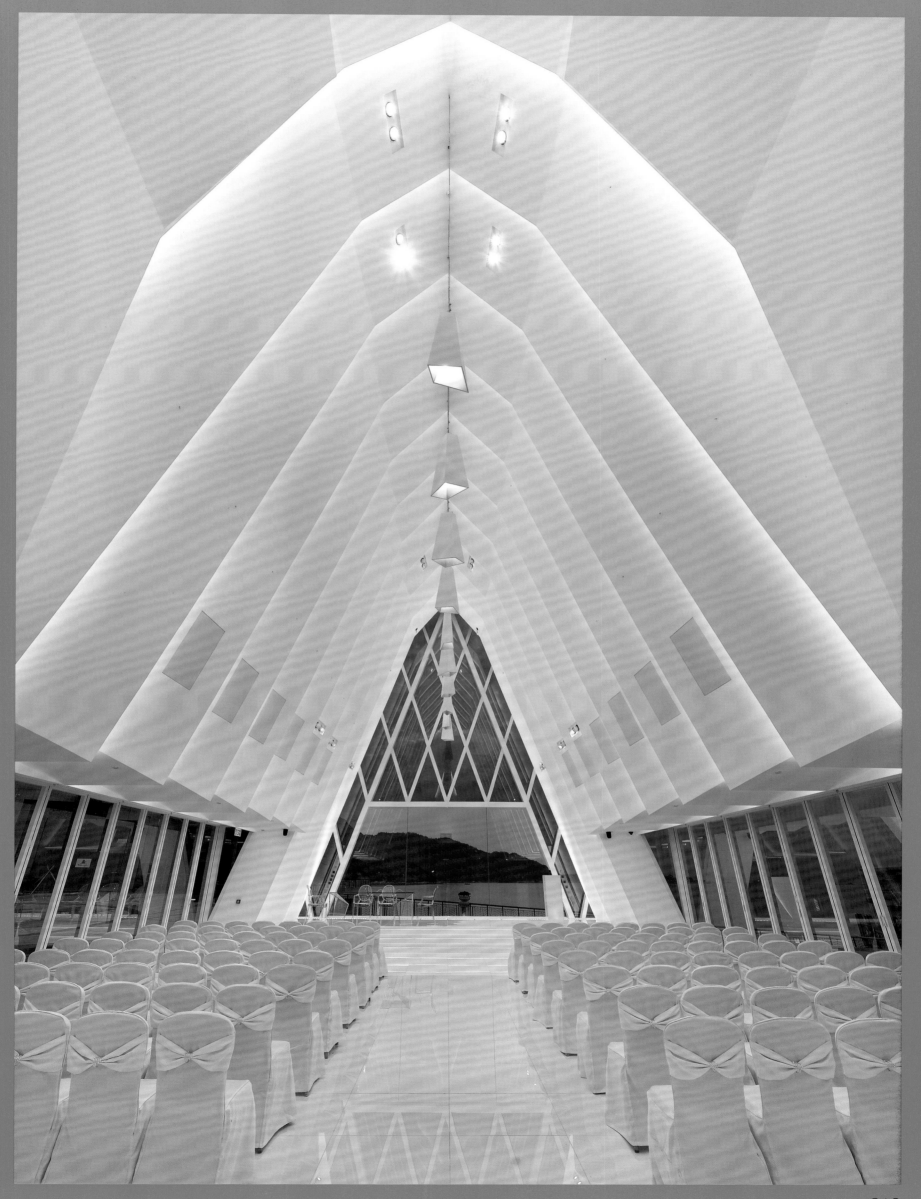

- PUBLIC SPACE
  公共空间
- BRONZE MEDAL
  铜奖

DALIAN INTERNATIONAL CONFERENCE CENTER

大连国际会议中心

Design Agency / Jiang & Associates Design

In the case of interior design in fully extended Austrian Coop Himmelblau under the premise of building design concept, to the interior space of modeling, color, material texture analysis such as restructuring, so as to make it flow fully reveals the contemporary sense of architectural space, feeling, science and technology feeling and openness. Used in the design of a large number of streamline modelling, applied on the modelling of metope, smallpox, hyperbolic form in modern before a technique in this paper, the Dalian sea split the beauty of the landscape. The design of the project closely the theme of green, environmental protection and energy saving, people-oriented.

本案室内设计在充分延续奥地利蓝天组建筑设计理念的前提下，对室内空间的造型、色彩、材料质感等进行分析重构，使其充分展现建筑空间的现代感、流动感、科技感及开放性。设计中运用大量流线造型，在墙面、天花的造型上应用双曲形态，以此现代前卫的手法阐述大连依山傍海的美丽风貌。项目的设计紧扣绿色、环保节能、以人为本的主题。

Design Team / Frank Jiang, Chen Wentao, Qin Gang
Location / China
Area / 140,000 m²
Client / Bureau of Urban Planning, Dalian, China

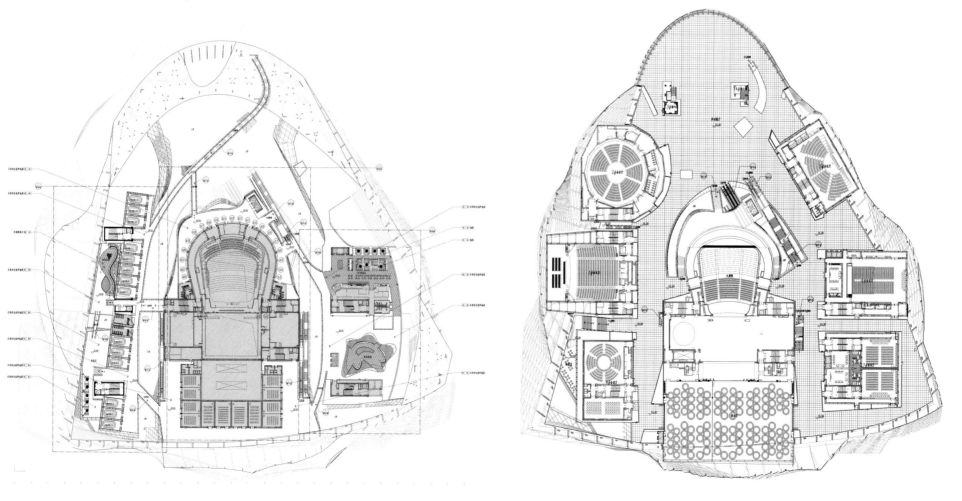

CONFERENCE CENTER

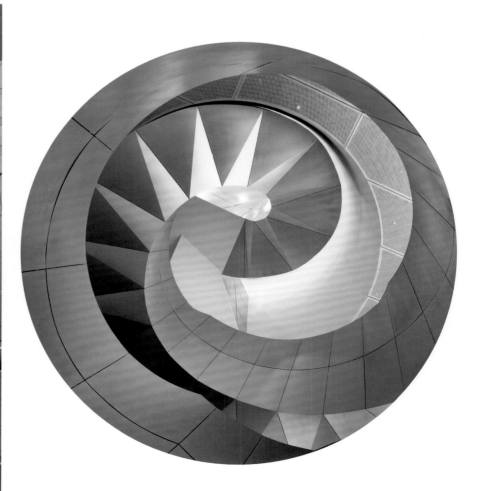

CONFERENCE CENTER

● PUBLIC SPACE 公共空间   EXCELLENCE MEDAL 优秀奖

CHONGQING VANKE YUEWAN SALES CENTER

重庆万科悦湾销售中心

Design Agency / Shenzhen Matrix Interior Design

The design of Yuewan Sales Center blends Asia elements in modern architectural language, combines traditional artistic conception with modern style and uses modern design to metaphorize Chinese tradition. Fraxinus mandshurica screen goes with dark stone, creating charming symmetry in the space. It creates a 3D space and at the same time demonstrates the classical elegant temperament of the east. ●

悦湾销售中心将亚洲元素植入现代建筑语系，将传统意境和现代风格对称运用，用现代设计来隐喻中国的传统。水曲柳屏风与深色石材的搭配既传统又流行，而且为空间营造出了充满魅力的对称感。整个空间具立体感，在美观之余，更增韵味，彰显东方的古典优雅气质。

Designer / Wang Guan
Location / China
Area / 1,100 m²
Client / Chongqing Vanke
Main Materials / Marmala White, Shell Mosaic, Marble, Metal Blinds, Black Steel Fluorocarbon Lacquer Surface, Black Mirror, Wood Veneer Washed White

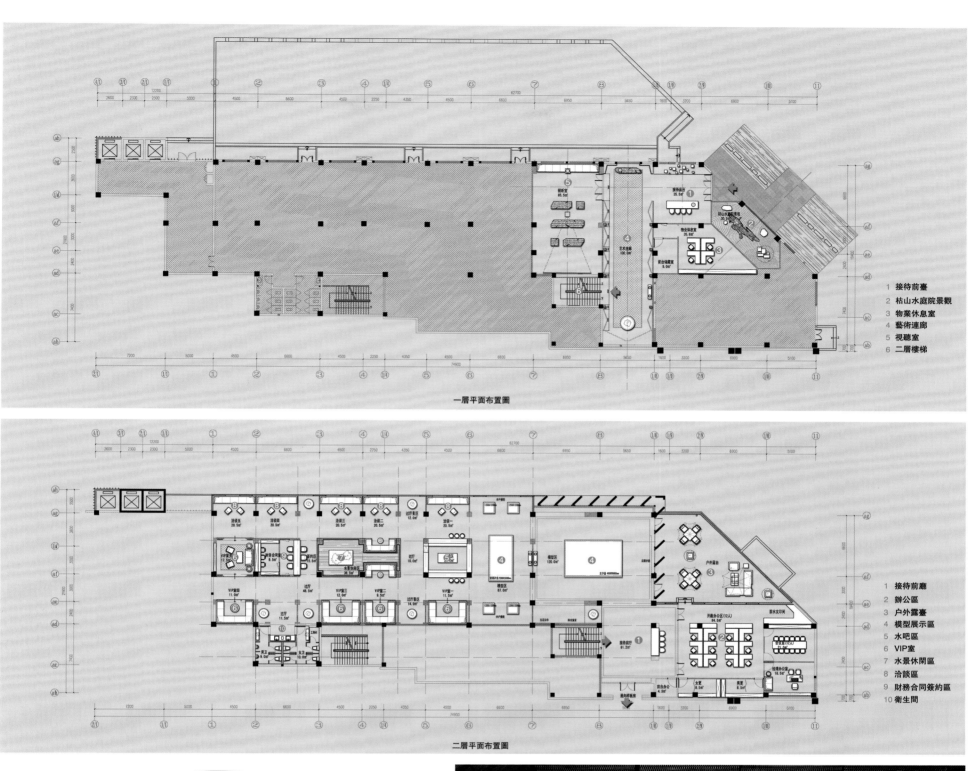

一層平面佈置圖

1 接待前臺
2 枯山水庭院景觀
3 物業休息室
4 藝術連廊
5 視聽室
6 二層樓梯

二層平面佈置圖

1 接待前廳
2 辦公區
3 戶外露臺
4 模型展示區
5 水吧區
6 VIP室
7 水景休閒區
8 洽談區
9 財務合同簽約區
10 衛生間

SALES CENTER

PUBLIC SPACE
公共空间

EXCELLENCE MEDAL
优秀奖

Design Team / Lyndon Neri, Rossana Hu and others
Location / China
Area / 2,400 m²
Client / Design Republic

DESIGN REPUBLIC DESIGN COMMUNE

设计共和设计公社

Design Agency / Neri&Hu Design and Research Office

Design Republic Design Commune, located in the center of Shanghai, envisions itself as a design hub, a gathering space for designers and design patrons alike to admire, ponder, exchange, learn, and consume. It houses the new flagship store for Design Republic, a modern furniture retailer, alongside a mixture of design-focused retail concepts, including books, fashion, lighting, accessories and flowers. The Commune will also have a design gallery, an event space, a café, a restaurant by Michelin-Starred Chef Jason Atherton, and a one-bedroom Design Republic apartment. ●

设计共和公社位于上海市中心，其定位是一个设计中心、一个聚集的场所，供设计师、顾客等人群来此观赏、沉思、交流、学习。它是设计共和——一个现代家具零售商最新的旗舰店，秉持混合的专注于设计的零售理念，售卖书籍、时尚品、照明、配饰和花卉等。其内还有一个设计画廊、一间活动室、一间咖啡厅、一间米其林星级厨师 Jason Atherton 的餐厅以及一间设计共和的单卧室公寓。

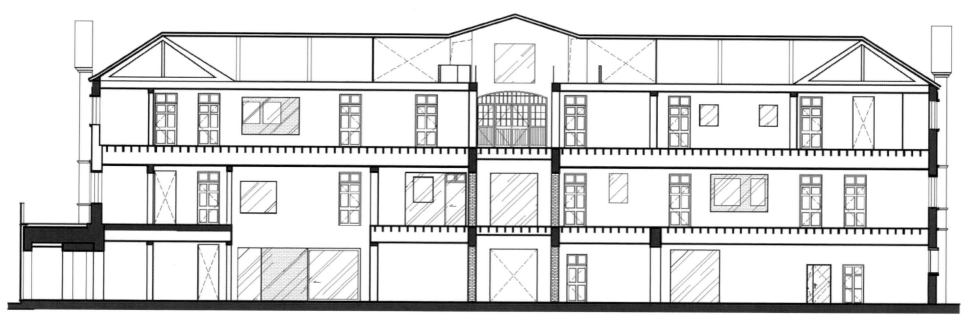

DESIGN COMMUNE

● PUBLIC SPACE 公共空间　　TOP TEN 十名入围

SPRING

春天

Design Agency / Joey Ho Design Limited

Spring is a learning center for children with the aim to help them achieve their development and learning potential during the crucial growth years. By bringing together the perspectives of a child and an adult, and manifested in the space, furniture and details, it is hoped to facilitate a dialogue and interaction between the children, parents and educators.

The reception counter is fitted with stairs to allow kids to climb up and communicate with the staff directly. The cooking studio is modeled after an adult's kitchen to give an authentic cooking experience. The tree houses and swings in the cafe are inspired by kid's playground, and a hilly relief feature create cozy cocoons for reading or internet surfing. Likewise, the bathroom features a dual height washbasin in the form of a fountain, which lets adults and children share the same facility and thereby forging a closer relationship. ●

本案是一个儿童学习中心，旨在帮助孩子们在最重要的成长期开发学习潜力。将儿童与成人的视角综合考虑，运用在空间、家具和细节中。希望能在儿童、家长和教师之间形成一种对话和交互作用。

接待柜台处配备了楼梯，供孩子们攀爬，与工作人员直接交流。烹饪室仿造成人厨房而建，提供真实的烹饪体验。树屋和咖啡馆内的秋千灵感来自于儿童游乐场，特色丘陵形成舒适的蚕茧，可以在此阅读或是上网。卫生间的特色是喷泉形式的双层高度的洗水盆，孩子和成人可共用此设施，有利于在他们之间培养更加亲近的关系。

Design Team / Joey Ho, Noel Chan
Location / Hong Kong, China
Area / 697 m²
Client / Spring Learning Limited

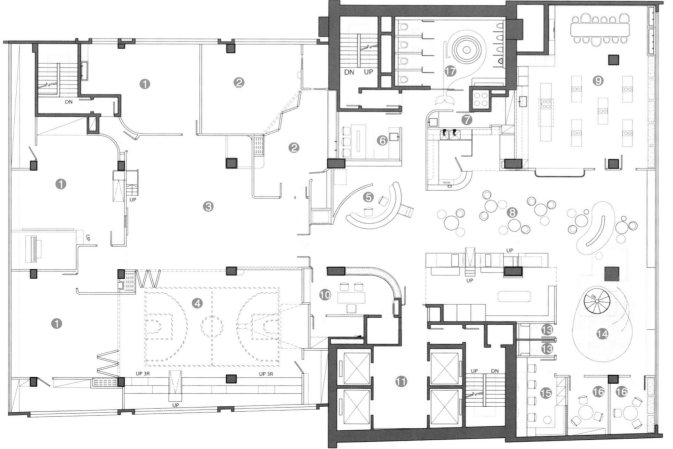

1	Kindyroo & Activity Room
2	Sensory Room
3	Kindyroo Apparatus Area
4	Multi-Purpose Sports Stadium
5	Reception Area
6	Pantry
7	Laundry
8	Cafe Area
9	Cooking Class
10	Principal Room
11	Lift Lobby
13	Feeding Room
14	Tree House
15	Staff Room
16	Meeting Room
17	Washroom

335

● PUBLIC SPACE
公共空间

TOP TEN
十名入围

Design Team / Maya Nishikori, Yushiro Arimoto
Location / Japan
Area / 320.49 m²
Client / Kentaro Masaki

G CLINIC

G 诊所

Design Agency / KORI architecture office & Arimoto Yushiro Architects Office

G Clinic is for hair restoration in Tokyo. The designers planned a universalistic arrangement by minimum handling for maximum effect with the growth of clinic.

On 8th floor, designers created a deep and continuous space for both men and women by sloping walls and furniture with same material, perforated MDF panels. Patients feel secure without worrying other patients.

On 7th floor, there is waiting room for many patients to consult and account. The voids on MDF panels are arranged as to control meeting of eyes of patients and staff. Void patterns are expanded on the sliding door, shelves and wallpaper. They give depth and sense of privacy.

9th floor is for an expanded counseling rooms and office. A diagonal stainless steel mirror wall guides patients smoothly from 7th floor to counseling rooms on 9th. And it expands the narrow space with a reflection from the diffusing light through the original fabric. ●

　本案是位于东京的一间头发护理治疗诊所。设计师计划用最少的设计处理来达成最大的效果，满足诊所的发展。

　在 8 楼上，穿孔的 MDF 板制成的倾斜墙面和相同材质的家具，创造出一个男女适用的有深度的连续空间。病人们会觉得受到保护，无需担心其他病人。

　7 楼是候诊室，可供许多病人咨询挂号。穿孔的 MDF 板被用作病人和员工之间的隔板。滑动门、架子和墙纸上都有孔洞图案，增加空间深度的同时保证一定的隐私。

　9 楼是扩充的咨询室和办公空间。成对角线的不锈钢镜面墙引导病人顺利地从 7 楼去往 9 楼的咨询室。镜面的反射让狭窄的空间显得更加宽敞。

- PUBLIC SPACE
  公共空间

TOP TEN
十名入围

Design Team / Angela Pang, Po Chung Yin, Thomas Yuen, Holman Chong, Dennis Chan
Location / Hong Kong, China
Area / 7,041 m²
Client / The Chinese University of Hong Kong

SPATIAL REORGANIZATION OF THE UNIVERSITY LIBRARY IN CUHK

香港中文大学图书馆空间改造

Design Agency / PangArchitect Limited in collaboration with Campus Development Office, The Chinese University of Hong Kong

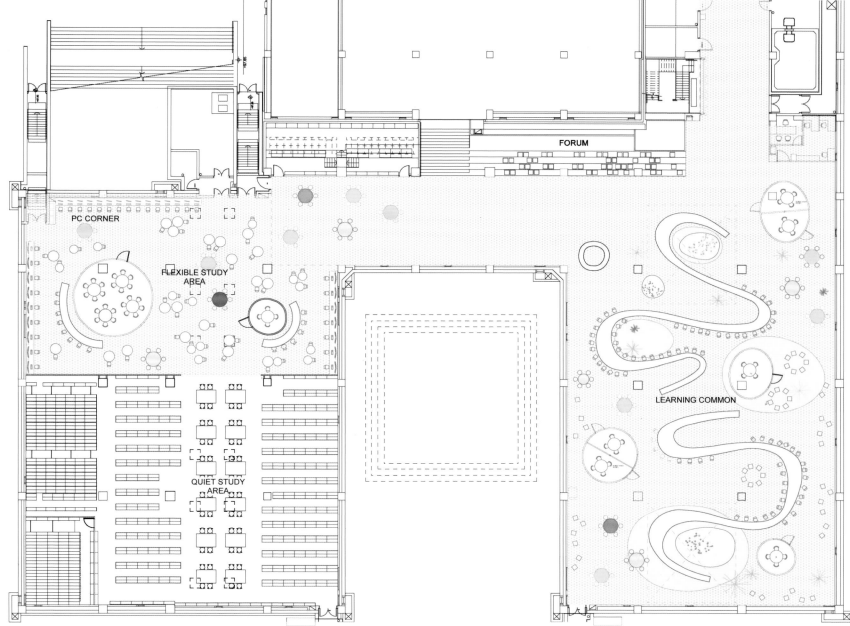

LGF

OF THE UNIVERSITY LIBRARY IN CUHK

This project aims to integrate the three buildings into one coherent whole to create an open and free plan where different learning activities can come together. Boundary walls were removed to enhance spatial interaction.

A central void is inserted into the library that brings light into what were once deep dark floor plates. Located within this void is the Grand Staircase. It is a physical and programmatic connection that brings together the interior. This openness of the interior encourages interactions among students for an active learning environment.

For the Learning Garden at LG/F, the long, sinuous table weaves students together through its curves and defines a series of zones. This ambiguity between public and private, openness and enclosure allows for different configurations of use. Subtle variations in the height, width, and shape of the Learning Path provide a datum line for students to discover new ways of learning. ●

本案设计目标是将三栋建筑合为一体，创造出一个自由开放的空间，鼓励学生来此进行各种学习活动。原有的分界墙被移除，以加强空间之间的互动。

图书馆中间设立了一个中庭，将自然光引入空间内。中庭内设置了大台阶，成为空间和功

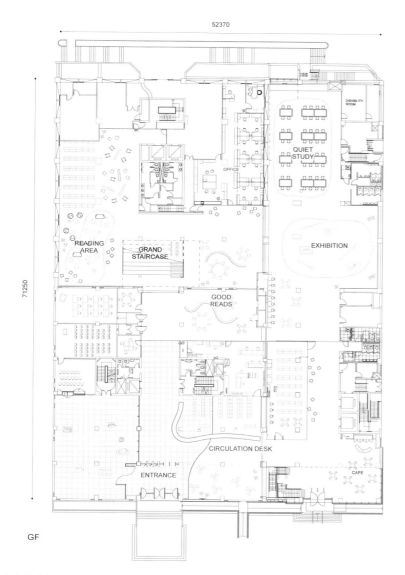 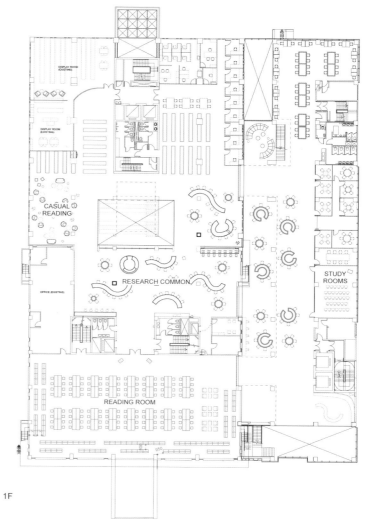

能上的连接,将室内空间联系到一起。开放的室内空间创造出活跃的学习环境,鼓励学生们进行互动。

在自修区放置了蜿蜒的长桌,将学生们联系到一起,同时通过其曲线划分出一系列学习区域,模糊了公共和私人界限的同时满足不同人的需求。弯曲的台面,其各处高度、宽度和形态有所不同,交织在整个空间中,让学生可以去发现和选择全新的学习方式。

- PUBLIC SPACE
公共空间

TOP TEN
十名入围

Design Team / YEN Bill CHEN-HSUN, Kellin Chen, Akon Li
Location / China
Area / 3,300 m²
Client / Greenland Group

GREENLAND M SALES CENTER

绿地•M 中心售楼处

Design Agency / Shanghai MRT Design Co., Ltd.

In the interior space of the sales center, designers adopt stair-step elevate gradually. There are ambulatories at special height. The whole space is similar to a big karst cave. At different heights, shining stalactite brings visitors special experience. Stairs with different size and height connect to each other, naturally join the space together. This is the first time designers try to isolate model area, negotiation area and bar area layer by layer. Glow wall is the highlight of the space. The whole space seems naturally generated by some kind of magic. ●

在售楼处的内部，设计师采用阶梯状逐渐抬高，在特别的高度采用回廊连接，整个空间似一个大的溶洞，可以在不同的高度体验闪闪发光的钟乳石带给参观者的特别感受。不同大小和高度的楼梯相互连接，空间自然流畅地连接在一起。设计师首次尝试将模型区、洽谈区、水吧区逐层分离又相互影响，发光墙是空间塑造的亮点，整个空间似通过某种魔力自然生成。

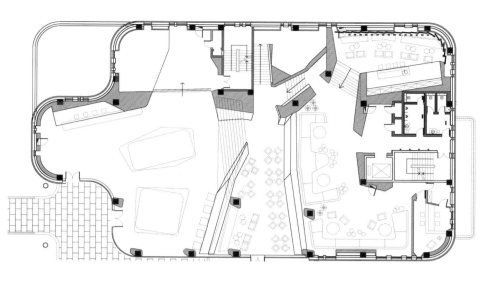
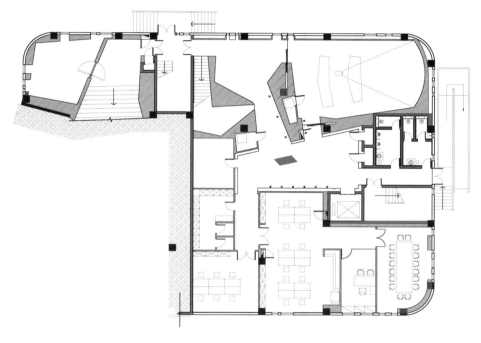

SALES CENTER 343

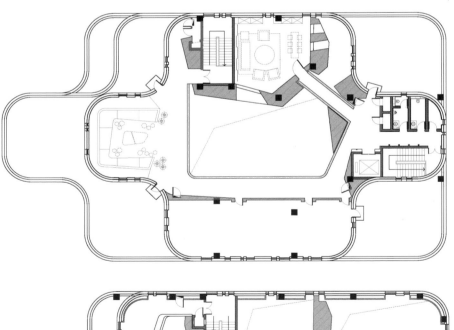
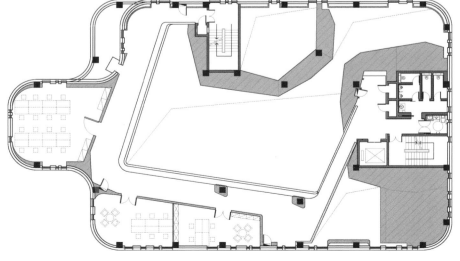

SALES CENTER 345

● PUBLIC SPACE
公共空间

TOP TEN
十名入围

ANHUI · ANQING FUCHUN ORIENT SALES CENTER

安徽·安庆富春东方销售中心

Design Agency / Shenzhen Matrix Interior Design

Designer / Cheng Jun
Location / China
Area / 1,200 m²
Client / Anhui Fuchun Real Estate Co., Ltd.
Main Materials / marble, varnish, white oak wood veneer, black mirror

Designer considers the integrality of the building and the interior space, and uses clean components in the space. Designer uses wood grain stone and carving black steel plate to create a natural and modern space. In terms of the decorations, improved table lamp and candlestick and delicate pottery connect the space as a whole.

安徽安庆富春东方销售中心,在设计上考虑建筑与室内空间的整体性,用干净简单的块面来处理空间,古木纹石材与黑钢板的雕花的应用,使得空间氛围朴实,有文化又有现代感。软装配饰上,经过改良的台灯及烛台、精致的陶罐,将空间以叙述的方式串联起来。

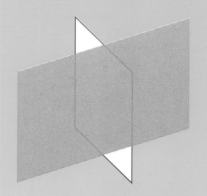

21st APIDA

WORK SPACE
办公空间

- **WORK SPACE**　办公空间
- **SILVER MEDAL**　银奖

Design Team / Hu Ruoyu, Zheng Chuanlu, Zhu Luxin
Location / China
Area / 1,000 m²

XIAMEN HIMALAYA DESIGN CO., LTD. OFFICE

厦门喜玛拉雅设计装修有限公司办公室

Design Agency / Xiamen Himalaya Design Co., Ltd.

The "Wooden Box" seemed like emerging from "nowhere", playing a major part in the overall space. The shadow of light, hanging of the rattans and the sloping wooden stairs danced along the way through open gaps between the "Wooden Box" and the plain white walls. Bunches of clay pots, brightened the expression of the Entrance Hall.

　　"木盒"四面临空，是空间的主角。光影、枯藤和错落的木阶梯在木盒和白墙间的缝隙中游走。地灯映射串串陶盆，丰富了门厅的表情。

The Reception Desk joined and connected the wood-line shadows on the concrete floor. The elevated pathway by the window appeared like a "catwalk" stage. The architect used the most common elements: concrete, raw clay, black steel, rusty iron, aged rattans and plain white wall, to create variable changes existing in one single space: tangible and abstract from near and far, up and down, wide and narrow, light and shadow…

　　木门虚掩，光影投射在水泥地及接待前台上。临窗走道架高，仿佛走秀的T台，穿透各个空间。设计使用了最常见的元素如水泥、粗陶、黑钢、锈铁、白墙等，最朴实的材料演绎了空间的无穷变化，虚实相间，远近结合，上下起伏，宽窄有致，光影跳跃……

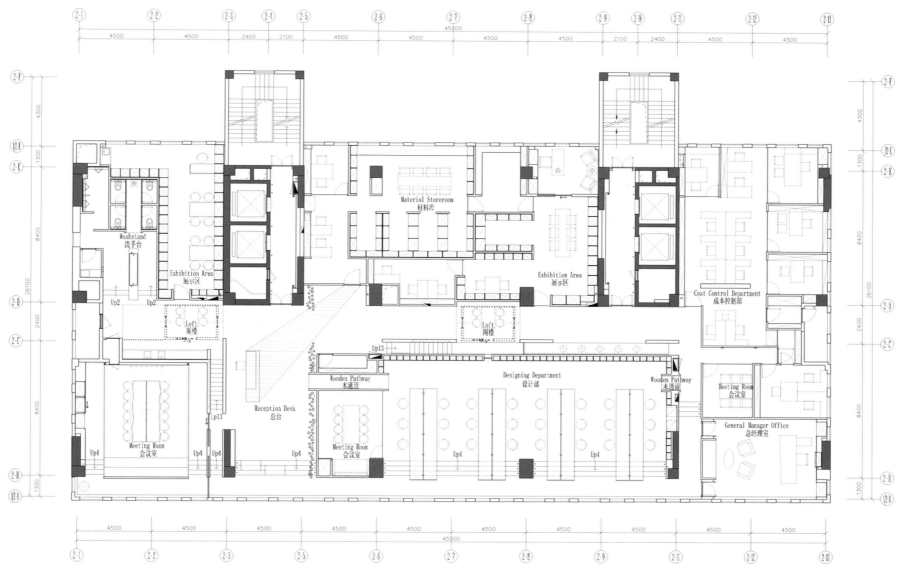

DESIGN CO., LTD. OFFICE

WORK SPACE 办公空间 BRONZE MEDAL 铜奖

THE MOMENT OF CHANGE

改变的时刻

Design Agency / YO Design Limited

Around the core of the triangle to the aisle, both the loops are connected to extend the space planning in all directions. Incorporate client's visual identity "The Beam" and the corporate color palette — yellow and grey as the main colors.

The floor configured with multiple large and small conference rooms and a four-in-one multi-purpose hall. There is a highlighted feature around the triangular core barrel. Design with a great variety of three-dimensional wall of dark wood in triangles of different sizes, criss-crossing the natural wood grain, forming a dramatic large-scale sculpture, yet solemn atmosphere full of artistic beauty.

Reception hall with white marble, suspended ceiling with the yellow logo extends out boldly exaggerated light trough, contrasting black circle rugs and cosy furniture. Glass is used as the main dividing wall, to help massive natural light into the space and enjoy the Victoria Harbour.

Design Team / Lam Cham Yuen, Olivia Lee, Jeffrey Choi, Daniel Lam
Location / Hong Kong, China
Area / 12,000 m²

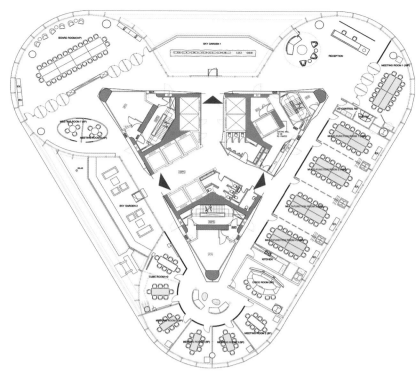

从核心三角到走廊，圆环形连接从各个方位延伸了空间的规划。设计将客户的视觉识别形象"横梁"与企业标准色结合，黄色和灰色是空间的主色调。

空间内配备了多个大小不一的会议室和一个四合一的多功能厅。三角形核心圆筒周围有一个非常突出的特点。由大量不同尺寸的深色三角形木块组成的 3D 墙，交叉着天然的木质纹理，形成戏剧化的大型雕塑，有着庄严的氛围又充满艺术魅力。

接待大厅使用了白色的大理石，悬挂的天花板上黄色的 LOGO 延伸出大胆夸张的光带槽，黑色的圆毯和温馨的家具形成鲜明对比。玻璃被用作空间的主要隔断墙，使得充足的自然光能进入空间内，同时将维多利亚港的美景尽收眼底。

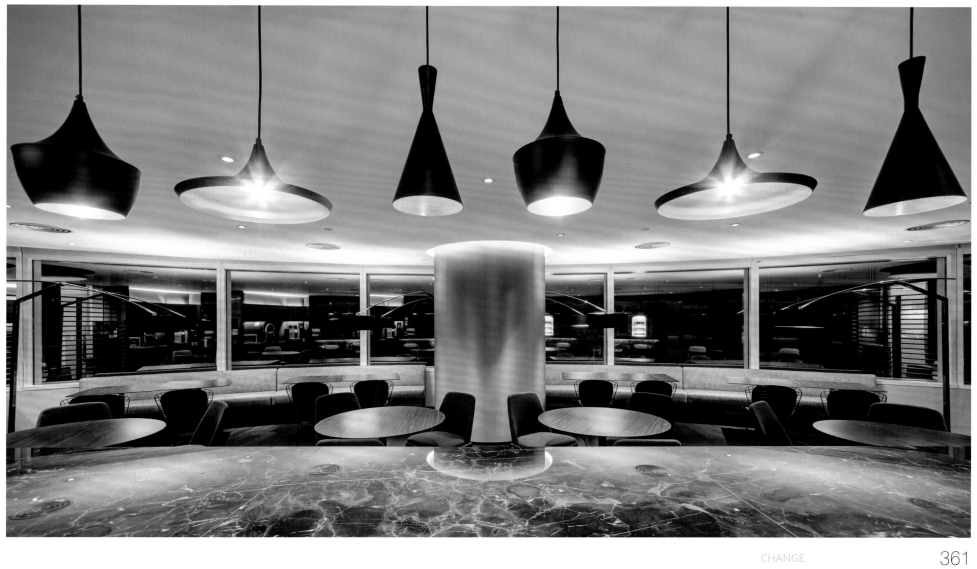

CHANGE

WORK SPACE 办公空间   **EXCELLENCE MEDAL** 优秀奖

MASAN SINGAPORE OFFICE

马山集团新加坡办公室

Design Agency / M Moser Associates

Masan Group's first office outside Vietnam is a symbol of the firm's rising international prominence, its future, and its deep roots in Asian tradition.

Most of the space is devoted to creating an evocative client experience. A film of water covers two-thirds of reception's floor area, with a conference room atop a wooden "raft" floating on its surface. Painstakingly engineered, the composition refers both to Vietnam's Mekong River and the office's Marina Bay location. Masan logos break the water's surface like stepping stones.

Transparent glass was used for all partitioning within the office, with a degree of privacy provided by leather panels — each repeating the Masan logo — at approximately eye-level. Furniture was selected to achieve a balance of contemporary aesthetics and a "residential" or "hospitality" feel that would soften the hard edges of the minimalistic architecture.

Design Team / Ziggy Bautista, Nirmala Srinivasa, Eliza Reyes, Adam Bentley, Chris Yeo, John Chen, Kenneth Chiam, Kim Horng Teng
Location / Singapore
Area / 464 m²
Client / Masan Group

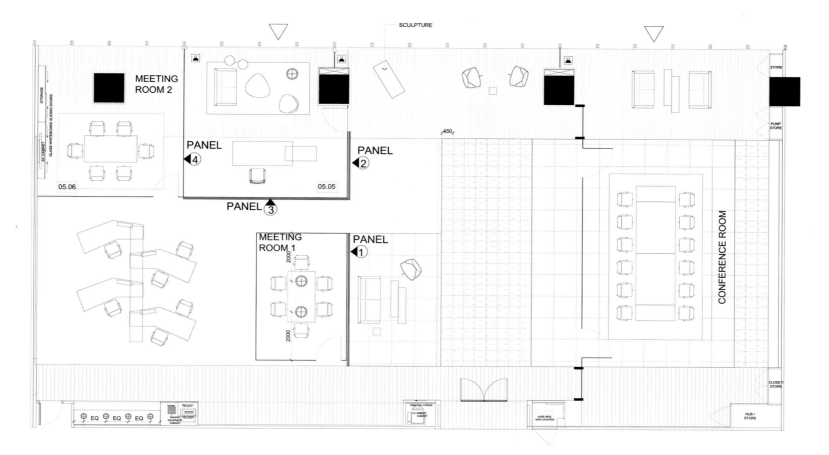

　　本案是马山集团在越南地区以外的第一间办公室，标志着公司日渐上升的国际地位、集团的未来及其深深的亚洲传统根源。

　　大部分空间被用来为来访的客户创造独特体验。接待层地面三分之二的面积都覆盖了一层水幕，会议室顶部漂浮着一个木制的筏子。这些煞费苦心的工程设计是为了指代越南的湄公河和办公室所处的滨海湾位置。马山集团的 LOGO 如同河面突出的石块一般，出现在水幕表层。

　　办公室内运用了透明的玻璃作为空间隔断，皮质的面板提供了一定程度的私密性，每块面板在接近视平线的位置都印有马山集团的 LOGO。家具的选用在现代美学和"住宅"或"接待"的感觉之间取得了平衡，柔和了简约建筑生硬的线条。

- WORK SPACE 办公空间
- EXCELLENCE MEDAL 优秀奖

Design Team / Tetsutaro NISHIDA, Mitsuhiko IMAI
Location / Japan
Area / 1,234 m²
Client / Sumitomo Mitsui Banking Corporation (SMBC)
Photography / Daisuke SHIMA (Nacása & Partners)

SUMITOMO MITSUI BANKING CORPORATION (SMBC) SHUKUGAWA BRANCH

三井住友银行夙川分行

Design Agency / NIKKEN SPACE DESIGN LTD

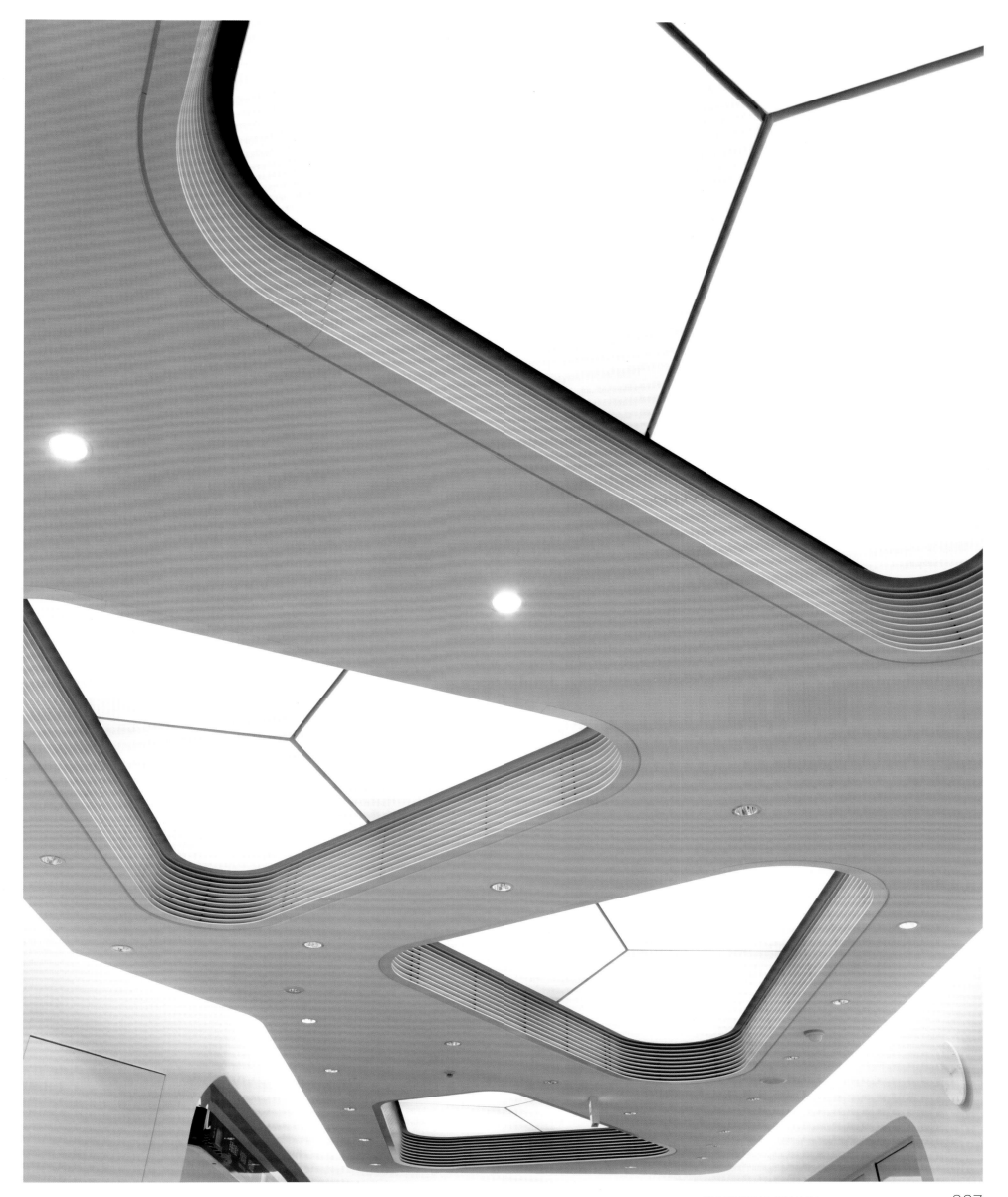

SUMITOMO MITSUI BANKING CORPORATION (SMBC)

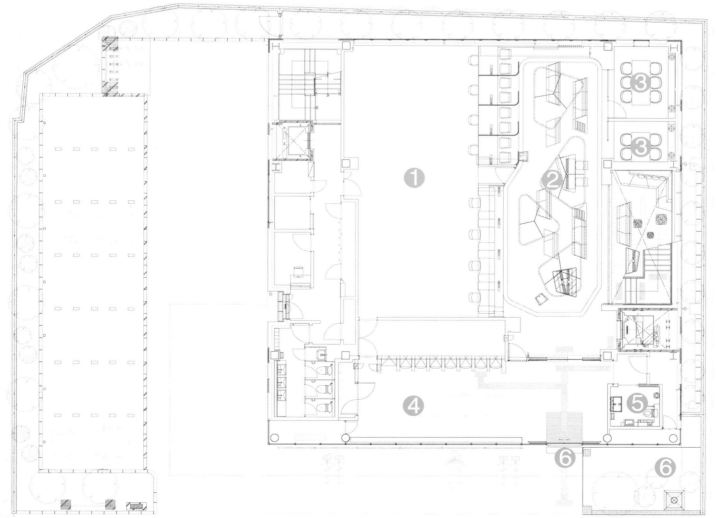

1. Office
2. Lobby
3. Reception room
4. ATM
5. Handicap toilet
6. Entrance

Ground Floor

The Shukugawa Branch of Sumitomo Mitsui Banking Corporation is located in a calm and beautiful historical town surrounded by nature forming a quiet venerable residential area representative of the Kansai region.

This branch is constructed of steel-framed reinforced concrete with a partially reinforced concrete structure, and has a basement and two floors above the ground. The building has a total floor area of 1,234 square meters and a service area of 472 square meters. Designers planned the design of the interior in order to create a cozy space that harmonizes with the daily lives of the local people, who are blessed with such a beautiful neighborhood. It is their hope that the branch will become a part of their lives, and provide the people and town with new opportunities and a new landscape.

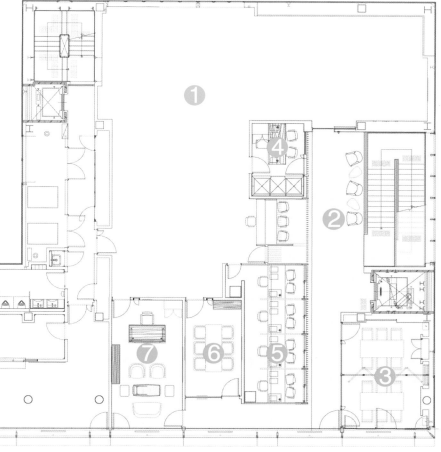

1. Office
2. Lobby
3. Convention room
4. Consulting room
5. Consulting booth
6. Reception room
7. Manager room

First Floor

三井住友银行夙川分行位于一个宁静美丽的小镇上，周围是安静的居民区。

银行由钢筋混凝土建造而成，有一个地下室和两层地面建筑。建筑的整体楼面面积为 1 234 平方米，服务区面积为 472 平方米。本案的室内设计旨在创造出舒适的空间，与当地人的日常生活和谐一致，与美丽的周边环境相得益彰。设计师希望这间分行成为当地人生活的一部分，给当地带来新的机遇和新的景观。

| WORK SPACE | TOP TEN |
| 办公空间 | 十名入围 |

Design Team / Joey Ho, Joe So, Michael Lam
Location / Hong Kong, China
Area / 789 m²
Client / Speedmark Transportation Limited

SPEEDMARK

立通

Design Agency / Joey Ho Design Limited

"Connection" is the keyword behind the design of this international freight forwarder headquarters in Hong Kong. It is about bridging the people working in the office and on the work sites in order to nurture a good sense of belonging to the company; and relating the work space to the business nature and corporate values of the company.

Upon arrival, the staffs are greeted by a reception counter resembling a real container in corporate blue colour. And a runway connects the individual working docks together. In general office, a 20 meter long semi-open container provides document storage for day to day operation. The staff cafeteria takes on the setting of a warehouse. The back-lit sky ceiling mural in the conference rooms, and the airplane windows on the walls in the waiting lounge all conjures up an impression of a company which aspires to connect the world.

本案是一个国际货运在香港的总部，设计关键词是"连接"。将在办公室工作的人们和在场地工作的人们连接起来，以培养员工对公司的良好归属感；将工作场地和公司的商业性质以及企业价值连接起来。

一进门，首先迎接员工的是一个类似真正的集装箱的蓝色接待台。走道将各个单独的工作台连接到一起。大办公室内，一个20米长的半开放式集装箱用于储存日常业务的文件。员工自助餐厅位于仓库边。会议室内的背光式天花壁饰与等待休息室墙面上的飞机窗口结合在一起，营造出渴望与世界各地联系起来的公司印象。

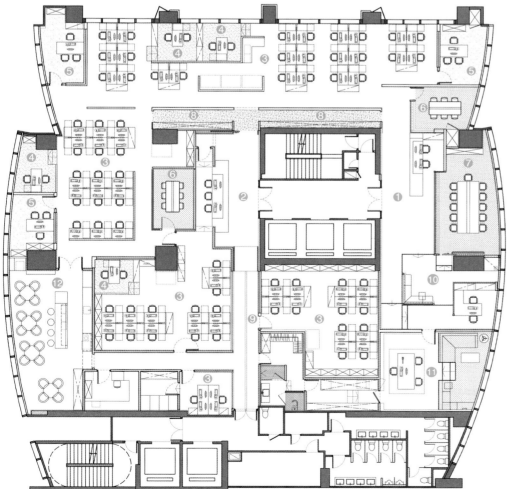

1. Guest Reception
2. Office Lobby
3. General Office
4. Team Leader Platform
5. GM Room
6. Meeting Room
7. Conference Room
8. Container Storage
9. Runway
10. Sky Lounge
11. District Manager Room
12. Warehouse Cafe

WORK SPACE 办公空间　　**TOP TEN** 十名入围

NANSHAN ZHONGTAI TIANCHENG OFFICE

南山中泰天成办公室

Design Agency / Shenzhen Matrix Interior Design

Nanshan Zhongtai Tiancheng office is used for sales office and sample office. The whole space is divided into two zones: sales area and sample office area. There is no obvious partition between zones. Two areas are both independent and connected with each other, thanks to the transparent glass partition. Sandwich layer makes the spatial level even more prominent; further improves the functionality and adds fun into the space.

　　南山中泰天成办公楼，主要功能为售楼与办公样板间展示。整个空间主要分为销售区和办公样板区两大块，两者间没有明显的隔墙划分，通透玻璃隔断让各区域相互独立又互相贯通。五米多的层高通过局部夹层的处理让空间层次更为突出，使用功能更为多样和富于趣味。

Location / China
Area / 612 m²
Main Materials / Beige Marble, Gold Marble, Diagonal Grain White Oak Veneer, Gold Toughened Glass, Black Steel Fluorocarbon Paint, Coated Glass, Green Carpet, Camel Carpet, etc.

| WORK SPACE | TOP TEN |
| 办公空间 | 十名入围 |

Design Team / Lam Cham Yuen, Olivia Lee Ming Yan
Location / China
Area / 2,300 m²
Client / Golden Concord Holdings Limited

GCL SHANGHAI HEADQUARTER

GCL 上海总部

Design Agency / YO Design Limited

Visitor feels immediately a bright environment once steping into the entrance lobby.

At reception area, while keeping the main logo wall aside, the stunning spectacular Pudong view can be introduced into the reception area. The designer uses the creamy marble and carpet flooring as the major material contrasting with the special dark wood veneer-Chickenwing wood. The whole environment looks sophisticated and modern, but yet a Chinese sense of belonging. Automatic sliding door leads to the dining room next to waiting area. The board room is mainly in darker wood finishes and browny leather, outstanding by the light cushion carpet and creamy upholstery wall panels. Cafeteria space is more welcoming and relax, forming a bar counter and bench seating.

Enclosed offices are allocating against the core wall while general staffs area are at the window side, to give natural light and achieve a transparent productive working environment in a Chinese culture company.

步入前厅，呈现在访客面前的是一个明亮的环境。

接待区除了有公司LOGO的墙外，美丽壮观的黄浦江景色亦扑面而来。设计师使用了奶油色大理石和地毯作为主要材料，与深色的木饰板形成对比。整体环境显得精致而现代，同时具有中式风格。自动滑动门通往紧挨着等候区的餐厅。董事会议室的设计主要使用了深色木材表层和深褐色的皮革，浅色的地毯和奶油色的墙板软垫给空间增色不少。自助餐厅则更为舒适放松，由长吧台和长条座椅组成。

封闭的办公室位于核心墙的后面，普通员工办公区则位于窗户旁，拥有良好的采光和通透的办公环境。

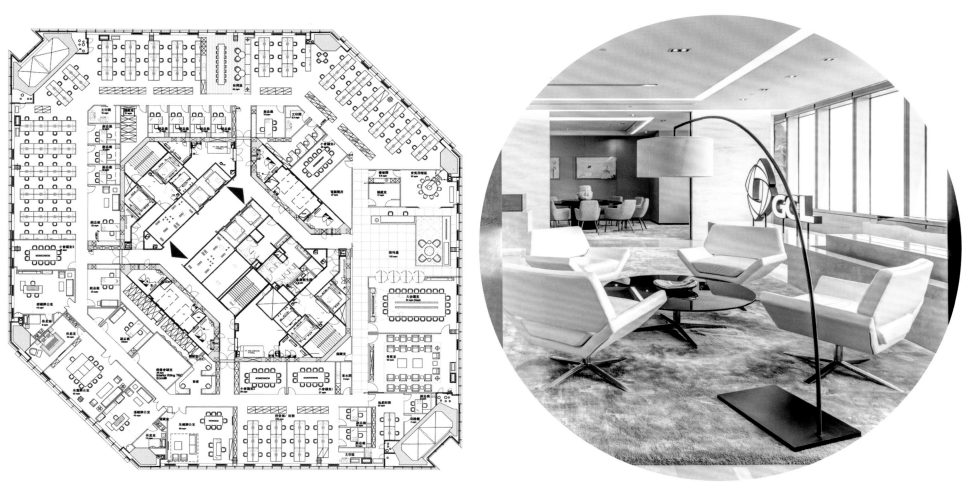

- WORK SPACE 办公空间
- TOP TEN 十名入围

XIAMEN FENGYA DESIGN CO., LTD. OFFICE

厦门风亚建筑设计顾问有限公司办公室

Design Agency / Xiamen Fengya Design Co., Ltd.

White color combines with natural color of wood forming the major color scheme of the space. Feature wall built with log is used as the partition of corridor, reception room and meeting room. In the open working area, there is a white long table, making the manager workstation relatively independent, at the same time, creating easy and comfortable atmosphere.

Pure space combines with wood carving and sculptures from master artists expressing designers' pursue toward simple and natural office space with artistic atmosphere.

Design Team / Zhang Yihui, Zhang Shuxiang, Chen Aqing
Location / China
Area / 280 m²

整个空间以白色为主，搭配自然的木色。一道互相重叠的原木造型墙，作为走廊、会客室、会议室的空间区隔，既开放又独立。敞开办公区的一条白色的长吧，使主管位相对独立，又使原本比较严肃的办公区域变得轻松、休闲。

简单纯净的空间配以古今艺术大师的雕塑表达了设计师对纯朴、自然、更具艺术氛围办公空间的追求。

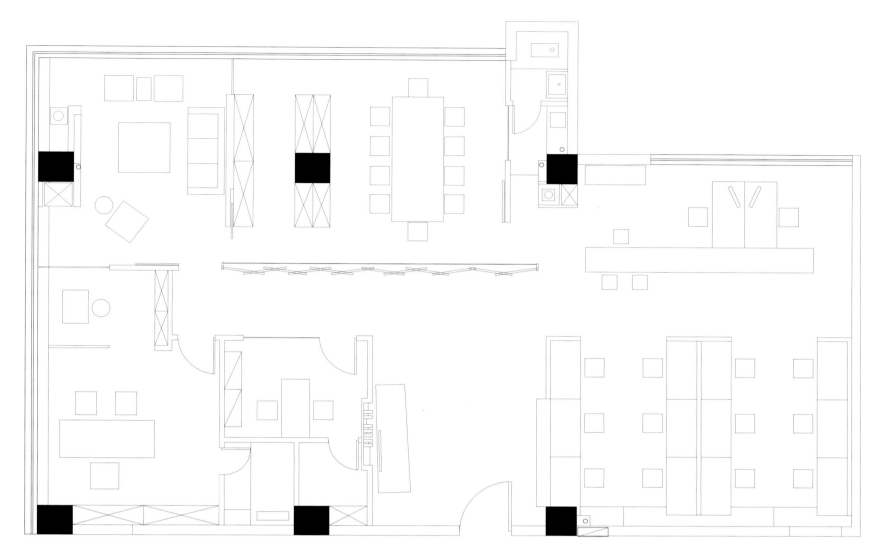

OFFICE 385

- WORK SPACE 办公空间　　TOP TEN 十名入围

Location / China
Area / 800 m²
Client / Fujian Huakun Investment Co., Ltd.

HUAKUN INVESTMENT

华坤投资

Design Agency / Fuzhou Lin Kaixin Interior Design Co., Ltd.

This project is an investment company which functions mainly as office areas for high-ranking leaders and reception areas for VIP customers. By bringing traditional elements and modern ones into each other's presence, this project seeks to find harmony in seemingly contradictions and try to illustrate oriental culture in a simplistic fashion, creating a relaxed and harmonious indoor space where "cultural aroma" and "sense of modernity" co-exist in perfect peace and fostering an atmosphere that nature and man in one. In architectural form, this project stresses the importance of catering to the aesthetics of the easterners in their obsession with magnificence and splendid sense of order, reflecting the robustness and steadiness of the enterprise while at the same time not losing sight of the innovative nature about the enterprise.

本案是一家投资公司，是一个以高层领导办公及VIP客户接待为主的场所。设计通过现代与传统的碰撞，从矛盾中找和谐，以简约手法诠释东方文化，营造一个"文化性"与"当代性"和谐并存的轻松、和谐的室内空间，营造天人合一的意境。在建筑形态上，强调符合东方人审美情感的建筑气势和庄严的秩序感，烘托企业稳健而不乏创新精神的特质。

一层平面布置图

二层平面布置图

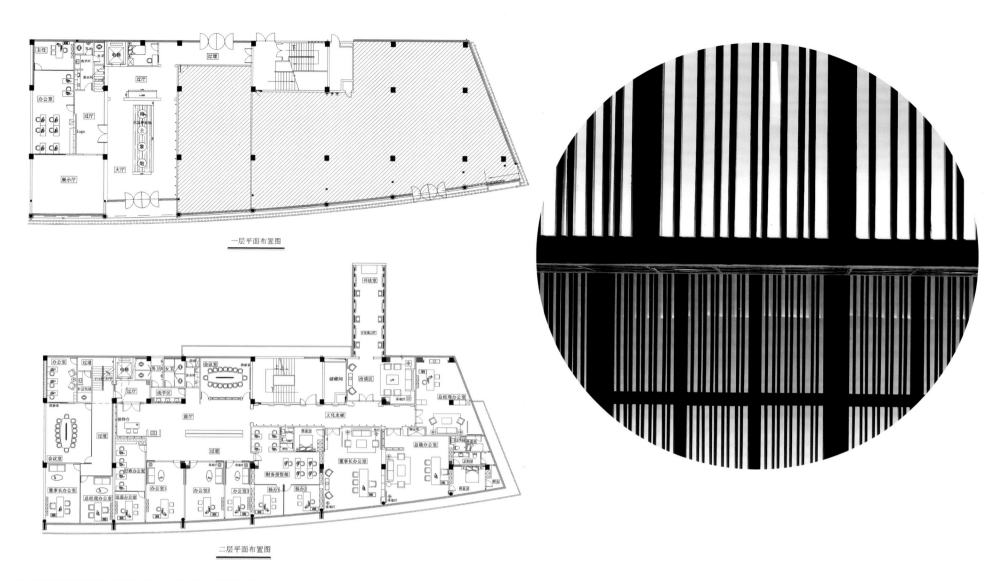

- WORK SPACE 办公空间
- TOP TEN 十名入围

MERCK SHARP & DOHME TAIPEI OFFICE

默克沙东台北办公室

Design Agency / M Moser Associates

The new Taipei office of Merck Sharp & Dohme (MSD) aims to accommodate a growing staff in a flexible, open, collaborative environment optimised to suit team and individual work needs.

To stimulate staff creativity whilst providing a visual cue to MSD's health-related business, a "natural" design theme prevails throughout the office. It is expressed both materially — through timber flooring, for example — and aesthetically through such features as patches of green lawn-like carpet and tree-like columns. The design also emphasises a good dispersal of natural light.

Work areas feature 120-degree workstations which encourage collaboration between staff members. Meeting spaces — both open and enclosed — are interspersed throughout the work area. Some are optimised for use by specific teams, such as the marketing department's cylindrical brainstorming room. Other striking features include a town hall space with cube seating, an oval staff pantry, and a wood-decked external terrace. ●

Design Team / JoanneChen, SnowyHsu, SophiaLo, SteveSze, Kuo C. K., Daniel Chen
Location / Taiwan, China
Area / 2,631 m²
Client / Merck Sharp & Dohme

& DOHME TAIPEI OFFICE

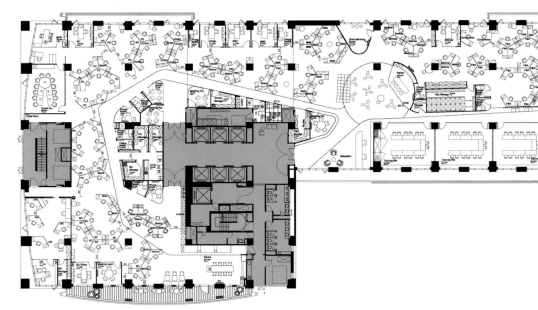

12F Furniture Layout Plan - R17

默克沙东全新的台北办公室的设计旨在创造出灵活、开放、合作的办公环境,适应公司不断增长的员工以及团队的工作需求。

为了激发员工的创造力同时展现MSD与健康相关的业务形象,"自然"的主题贯穿本案的空间设计。从材料上来说使用了木地板;从美学上来说,应用了一块块如同绿色草皮一般的地毯和树状的柱子。设计同样强调了良好的自然采光。

工作区域的特色是120度的工作站,鼓励员工之间的合作。会议室分散在工作区域中,有开放式的也有闭合式的。其中有一些是为特定的团队设计的,例如为市场营销部门设计的圆柱形集思广益会议室。其他一些突出的设计包括配备立方体座椅的大厅,椭圆形的员工餐厅和有着木甲板的外部露台。

WORK SPACE 办公空间　　TOP TEN 十名入围

Chief Designers / Huang Yongcai, Cai Lidong
Location / China
Area / 480 m²
Main Materials / Chinese Ash (fraxinus mandshurica), enhanced composite wood floor, ultra clear glass, frosted stainless steel.
Photography / Xiao Maoquan

SAMLEE OFFICE

仕其商贸有限公司（台湾、香港）广州办公总部

Design Agency / HONGKONG REPUBLICAN URBAN DESIGN CO., LIMITED

Without fussy details, the Samlee Office was designed by a simplicity oriental aesthetics. This concept matches with the speedy developing city. In this highly running information society, the project presents the interactive relationship between the city, work and people — a kind of intimately relation of activity and inertia; transparent overlay; permeation blank.

This project is a result of extending minimalism from the city metabolism. The "core line" just like a meandering river, the people flow in it, corresponds to the oriental aesthetics of "opening, developing, changing and concluding". The twist centre line through up all the function area.

A special "blank-leaving", with the triangle transparent fold line, overlaying and infiltrating the space, build up a four- dimensional binary relations between people and time.

The main theory of this layout is running through the three original architect units. The concept breaks the original formalistic space, but assembles the public area, opened-up office, and semi-enclosed office organically. From dynamic to quiescence, the users move in a streamlined plane and fold lined elevation, presents the ambiguous and communicated interior space. ●

仕其的办公室主张去繁从简的东方美学工作方式，符合了高速发展的城市消费模式，在信息高度运转的当下社会，本案诠释了城市群体与工作互动以及个体间的关系：动与静，透明叠加，渗透留白的暧昧关系。

本案是从城市新陈代谢的极简主义引申的结果，其核心的"动线"如蜿蜒的河流，人"流淌"于

OFFICE 397

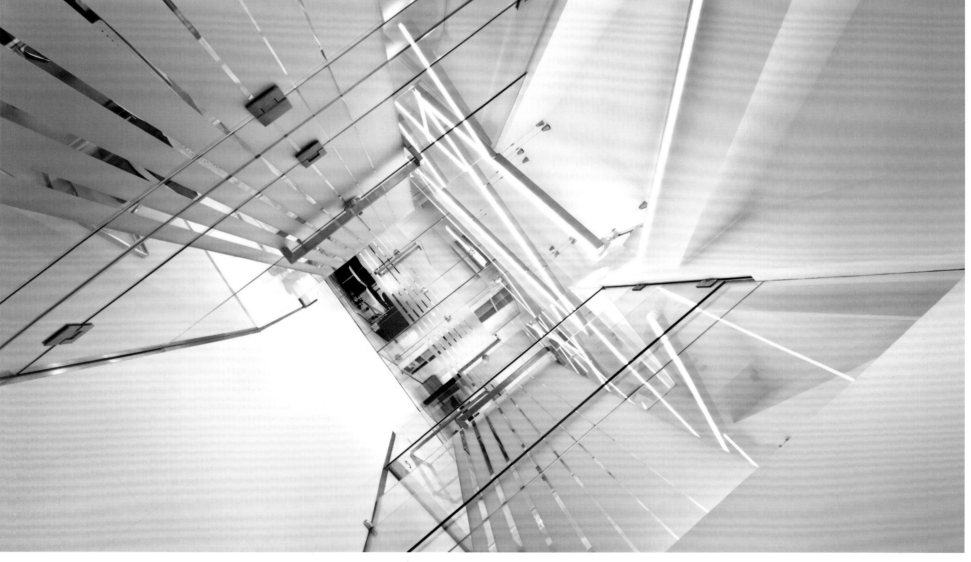

01.	入口/消防逃生口
02.	总台/前台
03.	样品室
04.	茶水间/休息室
05.	会议室
06.	洗手间
07.	女卫生间
08.	男卫生间
09.	洽谈室
10.	样品储存室
11.	机房
12.	公共办公区
13.	副经理室
14.	吧台/员工餐区
15.	文印室/员工茶水间
16.	董事长办公室
17.	总经理办公室
18.	财务室
19.	财务档案室
20.	经理办公室
21.	露台/吸烟区

动线正契合"起、承、转、合"之东方美学。曲折蜿蜒的动线贯穿整个功能与形式，其余部分"留白"，折线与三角形的透明、叠加、渗透模糊了空间界限关系，与人流动的时间形成四维上的对话成二元辩证关系。

在空间布局上其核心是贯穿原建筑三个单位的动线，把公共空间、敞开办公室区域到半封闭的办公空间有机组合在一起，打破原有的呆板空间布局。由动至静，人在动线与立面折线的叠加移动过程中形成空间的暧昧性、故事性。

家具平面布置图

21st APIDA
SHOPPING SPACE
购物空间

• SHOPPING SPACE 购物空间   SILVER MEDAL 银奖

INTERIOR DESIGN FOR TEA TAO (BROAD AND NARROW ALLEY STORE)

茶道室内设计

Design Agency / Alan Chan Design Company

The floor plan was divided into 3 zones to direct customer experience. The first part is the sales area. After that, customers enter a Zen-style courtyard where one can enjoy serenity. At the back of the shop, customers can taste the various products before making purchase decision. Instead of numerous small serving tables, a gigantic wooden table is used so that customers can mingle with each other.

Full-height glass is used to open up the whole store space. The courtyard and the tea tasting area can also be used for private events.

To deliver a natural feeling of Tea Tao, the materials used are natural and based on concept of 5 elements (metal, wood, water, fire, and earth) in ancient Chinese cosmology.

To enhance peaceful and sophisticated ambience, original design elements — Zen garden flooring, wall tiles with tea cup pattern, and acrylic bamboo with debossed Tea Tao theory is created.

Design Team / Alan Chan and Mr. Wong Kin-Ho
Location / China
Area / 80 m²
Client / Sichuan Emei-shan Zhuyeqing Tea Co., Ltd.

teaTAO - CHENGDU, CHINA

空间被划分成3个区域供顾客体验。第一部分是销售区。随后顾客可进入禅意庭院，享受宁静氛围。在做出购买决定之前，顾客可以在店面后方先行品尝各种产品。空间内，一个巨大的木质桌子取代了小小的服务台面，顾客可以互相交流。

落地玻璃让整个店面空间更加敞亮。禅意庭院及品茶区可以被用作私人活动场所。

为了表达茶道的自然意蕴，店内的设计选用了自然的基于中国古代的一种物质观——五行（金、木、水、火、土）的材料。

为了强调宁静和精致的氛围，设计师应用了原创的设计元素：禅意庭院地板、印有茶杯图案的墙面瓷砖以及刻有茶道理论的竹子。

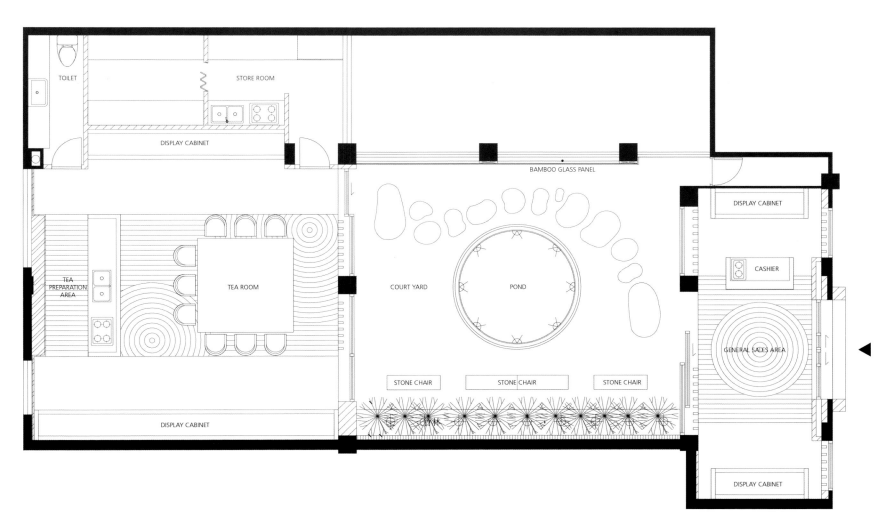

(BROAD AND NARROW ALLEY STORE)

- SHOPPING SPACE 购物空间
- BRONZE MEDAL 铜奖

ESTNATION FUKUOKA

ESTNATION 福冈店

Design Agency / MOMENT

The brand is named according to the concept of East-Nation, in short, spread from the Eastern countries to the world. Therefore it is important to harmonize the modern design with Eastern work and materials. The layout planning is the core to start design. For example, there are many alleyways in Japanese old towns, and it made all the walking more attractive. The necessary rooms as fitting and storage rooms are firstly located around the center of the shop. Then, the space between walls and these rooms works like the alleyways. They actually obstructs the view as the customers can not look around the whole view of the shop. However this structure would make them curious and walk ahead deeper, then they find more goods. It is a Japanese way of not showing the whole at once. We can find more if we do it little by little.

Design Team / Hisaaki Hirawata, Tomohiro Watabe
Location / Japan
Area / 404 m²
Client / ESTNATION INC.

品牌的命名来自于其理念——"东方民族",简言之,从东方走向世界。因此,在现代设计和东方产品及材料之间取得和谐平衡非常重要。平面布局规划是设计的开端,也是核心所在。在日本的旧城区有许多小巷,这使得步行更具吸引力。装置和储藏室首先被安排在店铺的中心附近。墙面和这些房间之间的空间成为如同小巷一般的存在。它们实际上是一种阻塞,因为顾客不能看到店铺的整体。然而,这些结构能激发人的好奇心,引导他们走得更加深入,然后发现更多商品。不要一次性地展示所有,这是一种日本式的思考方式。一点一点地去发现,我们会发现更多。

● SHOPPING SPACE 购物空间　　EXCELLENCE MEDAL 优秀奖

UM COLLEZIONI FEMALE TOP FASHION MULTI-BRAND STORE

UM Collezioni 高端女装综合时装店

Design Agency / AS Design Service Limited

Designers compares fashion to the sea. With the extensive experience in searching and collecting renowned luxury brands among the sea of fashion, UM Collezioni Female Top Fashion Multi-brand Store always pick the most fashionable pieces, pack and bring them back in their "Suitcase". This makes UM Collezioni standing in the leading position as always.

Customers flowed the circulation of shop and they will find a part of the suitcase opened and closed, which can increase the curiosity from customers. Products are placed with the wave form island display unit, everyone feels relaxed feeling, also increased their interest in the product. The display walls unit shelves design idea from the suitcase with variability of display combination. Designers created a unique wave-shaped island display unit and had a consideration of aesthetics, proportion, structure, components, transportation, and material properties.

　　设计师将时尚比喻成大海。UM Collezioni 高端女装综合时装店从时尚的海洋中精心搜寻知名的奢华品牌，挑选最流行的时尚物品，将其放进"手提箱"收集回来。这使得 UM Collezioni 始终站在时尚的前沿。

　　顾客在店铺中穿行浏览，他们将会发现一些打开和闭合的手提箱，这就增加了顾客的好奇心。墙面陈列架的设计灵感来自于手提箱，可以组合出多种陈列形态。设计师创造出独特的波浪形岛状陈列组件，考虑到了美感、比例、结构、部件及材料的特征。

Design Team / Four Lau (Creative Director), Sam Sum (Art Director), Francis Wong (Interior Designer)
Location / Macau, China
Area / 126 m²
Client / World First Holdings

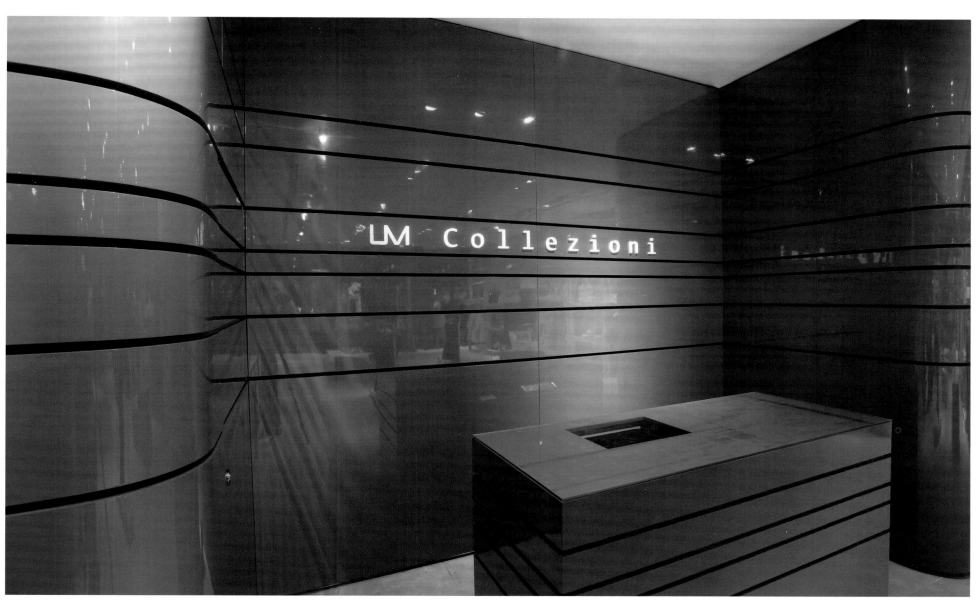

MULTI-BRAND STORE

● SHOPPING SPACE
购物空间

EXCELLENCE MEDAL
优秀奖

柏地广场

Design Agency / HHC DESIGN SOLUTION

BODHI VENUE department store is born due to the urgent needs of being "trendy & unique" and it uses design to narrate a consumption fable. Through the given theme on each floor, the combination of brands with different tastes together creates a novel sensual experience and a life theater with warmth.

Today, a commercial market that relates to our livelihood also continues to evolve to form a close tie to our constant changing life mode.

Therefore, by using a brand new spatial combination, BODHI VENUE interprets our understanding of happiness today! ●

　　BODHI VENUE 百货商场的诞生源于人们对于"时尚和独特"的迫切需求，用设计述说了一种消费理念。每层空间都有其特定主题，不同品味的品牌组合到一起，创造出一种新奇的购物体验。

　　如今，与人们生活息息相关的商品市场将继续其与人们日常情绪变化之间的密切关系。

　　因此，通过全新的空间组合，BODHI VENUE 诠释了设计师对于幸福的理解。

Design Team / Jacky Chang/ Jerry C. Dung/ Katie Yang/ Jim Dong/ Karen Hsu/ Phoebe Li/ Alfie Huang
Location / Taiwan, China
Area / 793 m²
Client / BODHI VENUE

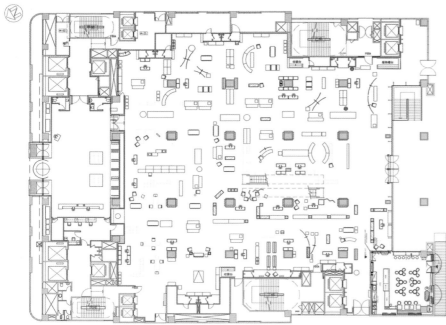

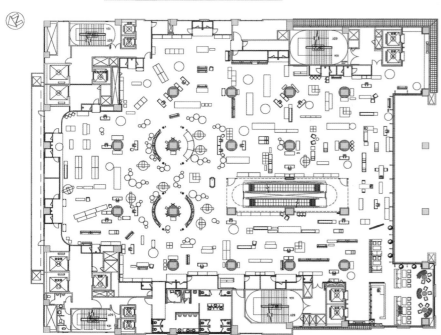

SHOPPING SPACE 购物空间 **TOP TEN** 十名入围

LOUNGE BY FRANCFRANC

Francfranc 休息室

Design Agency / A.N.D.

Lounge by Francfranc is a laboratory flagship shop which is newly developed by Francfranc with a concept of "casually stylish".

Designers covered an inorganic glass curtain wall with the illustration of European maison image. By separating the illustration drawn by an up-coming artist into two layers of black and white, and adding graphical arrangement, it has made the facade three dimensional and given an overwhelming presence.

Designers have succeeded in realizing an overpowering facade as a new landmark in a chic area, Minami-Aoyama.

Inside of the store, designers also used a strong contrast of black and white. By using various different materials, and expressing the texture quality hidden inside the colors, designers managed to give the depth into the space. It has become a space which informs a new brand identity by keeping the colors to the minimum and controlling a feeling of material quality.

　　Francfranc 休息室是 Francfranc 品牌旗下最新开张的一间实验性旗舰店，设计理念为"休闲时尚"。

　　设计师应用了无机玻璃幕墙，上面绘制着欧式住宅的图画。这幅图画由艺术家绘制而成，被分隔成黑白两个层次，同时增加了一些平面元素，让外墙呈现出 3D 的效果，具有强烈的视觉存在感。

Location / Japan
Area / 1,148.84 m²
Client / BALS Corporation

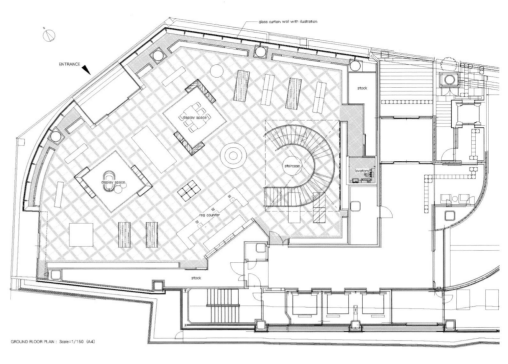

该外墙以其强烈存在感,成功地成为店面所在时尚地区的新地标。

店面内部的设计也运用了黑白的强烈对比。通过使用各种不同的材料,呈现出色彩中隐藏的纹理质感,给空间带来层次感。通过使用简约的色彩,控制材料所呈现的质感,空间表现出一种全新的品牌形象。

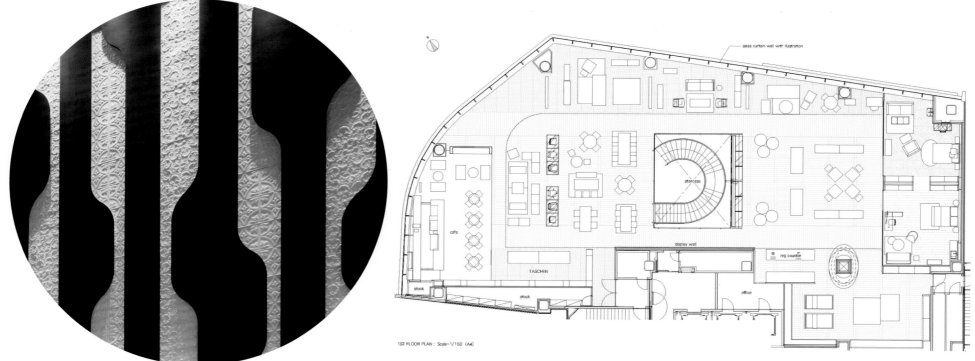

1ST FLOOR PLAN : Scale=1/150 (A4)

SHOPPING SPACE | TOP TEN
购物空间 | 十名入围

Designer / Stefano Tordiglione
Location / Hong Kong, China
Area / 85 m²
Client / apple & pie Ltd

APPLE & PIE

苹果 & 馅饼

Design Agency / Stefano Tordiglione Design Ltd

Playfulness, elegance and practicality come together at the new apple & pie boutique.

Inspired by the ethos behind the brand's name and its half apple-half pie logo, the concept combines the wellbeing elements represented by the apple with the more playful pie.

The former is reflected in the use of environmentally- and child-friendly materials with a focus on wood as opposed to plastic for the furnishings, while the latter can be seen in the whimsical interior design which ranges from bright red apple-shaped sofas to imaginative wall displays.

The seating hides storage space, while a lively tree design on one wall with white and red apples hanging from its branches, and elsewhere pie-shaped lattices and mounted fruit palettes, offer ideal shelving opportunities. In the windows, semi-circular pie-like features bring the logo and brand name full circle.

趣味、典雅和实用是全新的 apple & pie 专卖店的三个特色。

设计灵感来自于品牌名及其一半苹果一半馅饼的 LOGO，设计理念结合了苹果所代表的健康元素以及馅饼的玩味性。

前者体现在环境及儿童友好的材料的运用，主要是木材，与塑料的家具截然相反，后者则体现在一些异想天开的室内设计，包括明亮的红苹果形状沙发和富有想象力的墙面陈列。

座位下隐藏着储物空间，墙面上有着生动的树形设计，枝桠上挂满白色和红色的苹果，馅饼造型的格子和水果色搭配成为理想的货架。窗户处半圆形的馅饼状表现了品牌 LOGO 和品牌名。

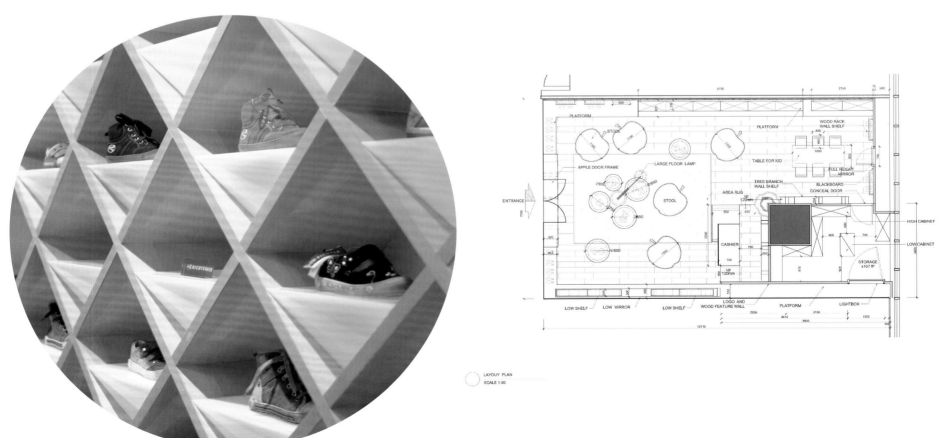

GUANGZHOU

● SHOPPING SPACE 购物空间   TOP TEN 十名入围

GUANGZHOU TEXTILE EXPO CENTER STORE

广州纺织博览中心商铺

Design Agency / 5+2 Design (Perceptron Group)

Location / China
Area / 310 m²
Client / Guangdong Dongcheng Liyi Group Co., Ltd.

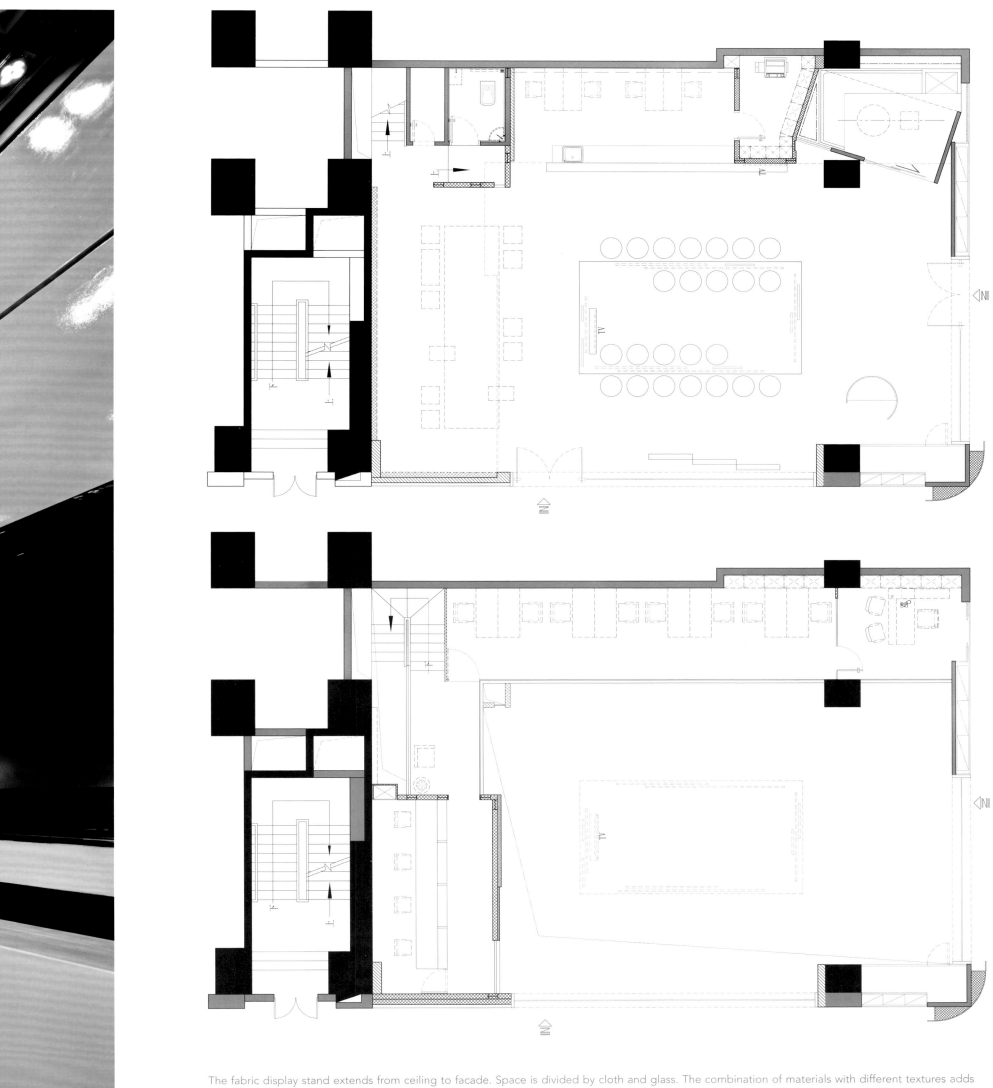

The fabric display stand extends from ceiling to facade. Space is divided by cloth and glass. The combination of materials with different textures adds vivid atmosphere to the space. LED lights with higher power are used according to different cloth and pattern in order to highlight the quality of products, creating remarkable shopping environment. Display shelves is flexible and can be grouped at will.

大厅中央的布艺展示台，从天花一直延续到立面，用布匹与玻璃划割出空间，突出了不同质感的材料之间的有序组合，为空间增添了跳跃的气氛。整个空间约有 6 米高，针对不同的布料、款式，使用较高功率的 LED 灯，产生清晰的色彩，很好地强调了展示物品的品质，创造引人注目的商店场景。一个个装置感很强的展架，令产品可以随意组合。

SHOPPING SPACE 购物空间　　TOP TEN 十名入围

Design Team / Pan Mok, Kimi Guan
Location / China
Area / 1,200 m²
Client / Takashimaya

MERCATO FRESCO

梅尔卡托弗雷斯科

Design Agency / 6070 Interior Design Ltd

Mercato Fresco is a Gourmet located at Takashimaya Basement associated with exotic fresh fruits, sundries, vegetables, fresh daily meat, high standard healthy food and all different kind of imported drinks.

The high-end yet cozy design looks appealing to the sophisticated customers. While the warm tone display units, high ceiling, pendant lamps and free hand drawings greatly enhance a comfortable shopping atmosphere for the discerning ones!

梅尔卡托弗雷斯科是一间美食购物广场，位于高岛屋底层，供应新鲜的水果、杂货、蔬菜、新鲜肉类、高标准健康食物和各种进口的饮品。

高档且温馨的室内设计富有吸引力。温暖色调的展示装置、高高的天花板、吊灯和手绘图案增强了舒适的购物氛围。

- SHOPPING SPACE 购物空间
- TOP TEN 十名入围

DFS LE SALON

DFS 沙龙

Design Agency / pmdl Architecture + Design

Situated within the DFS Galleria in the "Shoppes at Four Seasons" Macau, Le Salon is a unique retail environment showcasing some of the world's most exclusive and prestigious timepieces. The space embodies five environments: Entrance Foyer, Vintage Salon, Watchmaker's Salon, Accessories Salon and Private Consulting Lounge, giving the customer an intimate shopping experience.

The entrance reflects the modern, digital age with a three meter LED display recalling the dates and times of some of the most prestigious vintage timepieces in store.

Design Team / Simon Fallon, Vania Contreras, Gayatri Sawe, Andrew Pender
Location / Macau, China
Area / 217 m²
Client / DFS Cotai Limitada

The Vintage Salon is reminiscent of a turn-of-the-century museum, the vintage timepieces placed on individual displays with digital screens, displaying their provenance in a contemporary manner.

The diamond-shaped Watchmaker's Salon allows the customer upon entry to see all four of the carefully selected brands, whilst towards the end of the enfilade, the Accessories Salon displays some of the finest leather watch accessories.

本案位于澳门DFS风雨商业街廊内，提供了独特的零售环境，展示了一些世界最顶级的钟表产品。空间中展现了5种环境：进口门厅、古董钟表沙龙、钟表沙龙、配饰沙龙和私人咨询室，给顾客提供亲密的购物体验。

入口门厅代表了现代、数字化时代，通过一个三米的LED显示屏让人回忆起店内最珍贵的古董钟表所处的年代。

古董钟表沙龙让人联想到世纪之交的博物馆，古董钟表被展示在单独展架上，配备的数字显示屏展示了其渊源。

钻石形的钟表沙龙让顾客可以走入，浏览店内四种精心挑选的品牌，尽头处是配饰沙龙，展示了一些精美的皮革手表配饰。

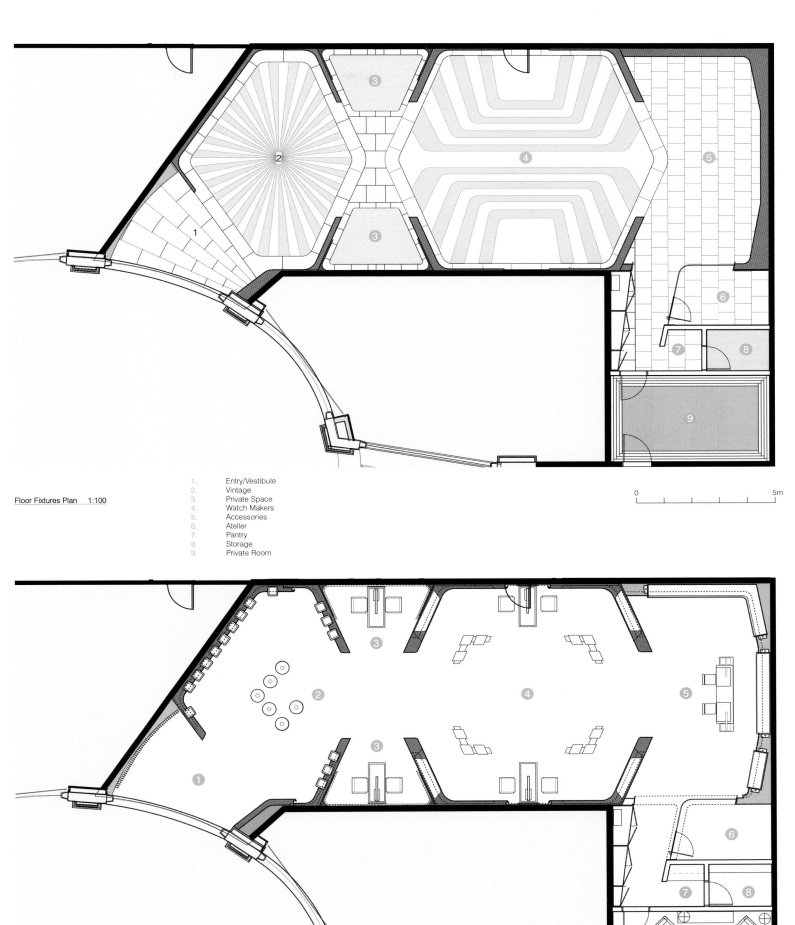

Floor Fixtures Plan 1:100

1. Entry/Vestibule
2. Vintage
3. Private Space
4. Watch Makers
5. Accessories
6. Atelier
7. Pantry
8. Storage
9. Private Room

SHOPPING SPACE 购物空间 **TOP TEN** 十名入围

Designer / Huang Sanxiu
Location / China
Area / 5,000 m²
Client / Langfang Defa Classical Furniture Co., Ltd.

LANGFANG DEFA CLASSICAL FURNITURE EXPERIENCE PAVILION

廊坊德发古典家具体验馆

Design Agency / Beijing KaTa International Architectural Consultation Co., Ltd.

Zen is dumbness but with infinite meaning and uncertain feeling. Everyone has his own understanding about Zen.

The quintessence of Zen is leave room for others to think about. In the space, virtuality combines with reality. You can see through the dark grilling on the white wall and glance at the rear scene, admiring delicate woodworks. Pieces of polished gentle woodworks standing there echo with each other, creating peaceful scenery.

Water, leaf, sand and stone create subtle landscape; give off a sense of Zen.

禅本无言，却有无限的意义与无常的感觉。每个人心中都有一个对禅的诠释。

禅的精髓在于"不说破"，留给他人思考的余地。空间的虚实结合，透过白墙上深色的格栅影影绰绰一瞥背后的光景，蜿游其中，欣赏木器之精美。一件件打磨温润的木器站在那里，互相遥望，似有低语，似有婉唱。

一泓水，一片叶，一粒沙，一枯石，细微之处游山水，悠然之间品禅心。

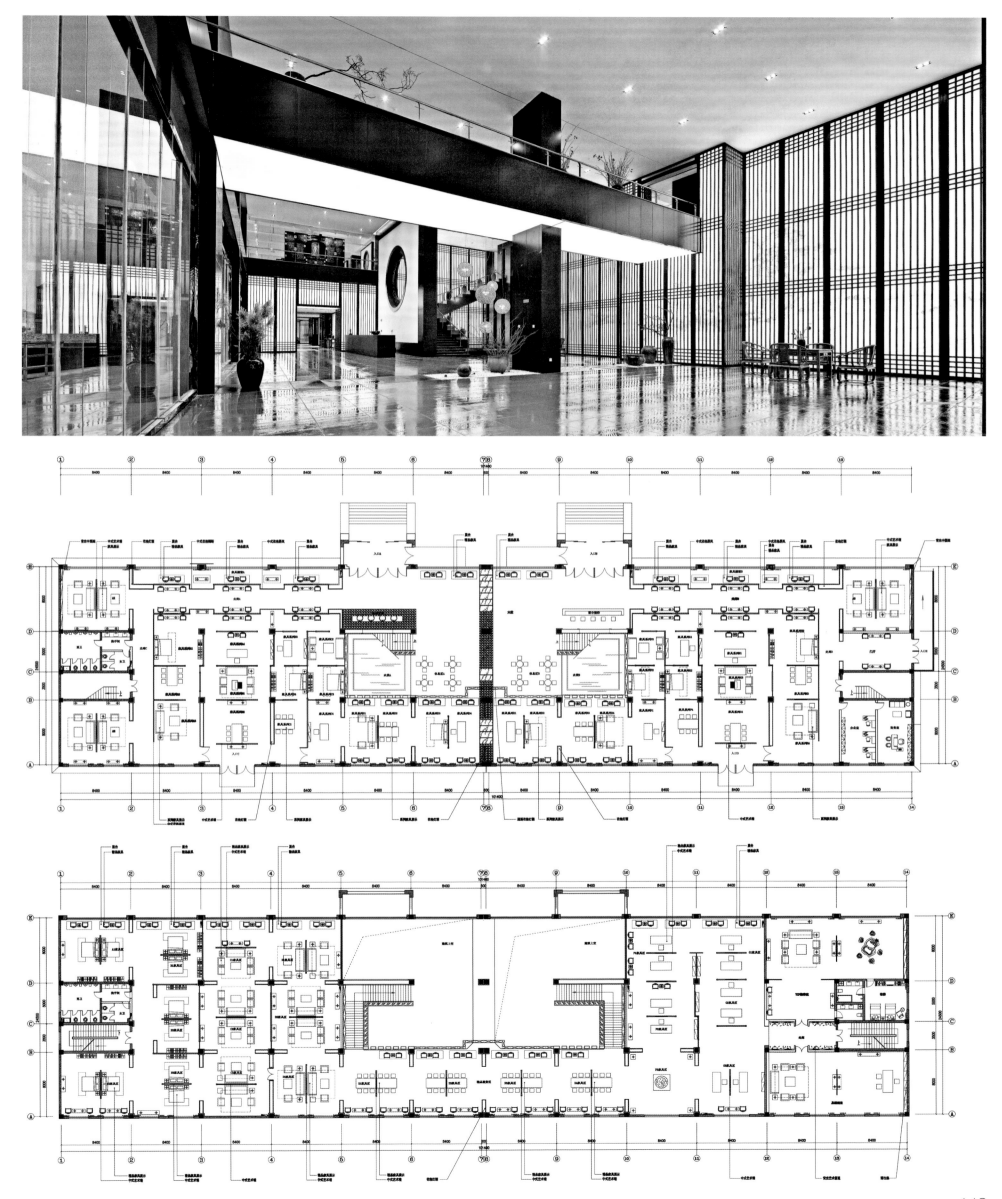

EXPERIENCE PAVILION

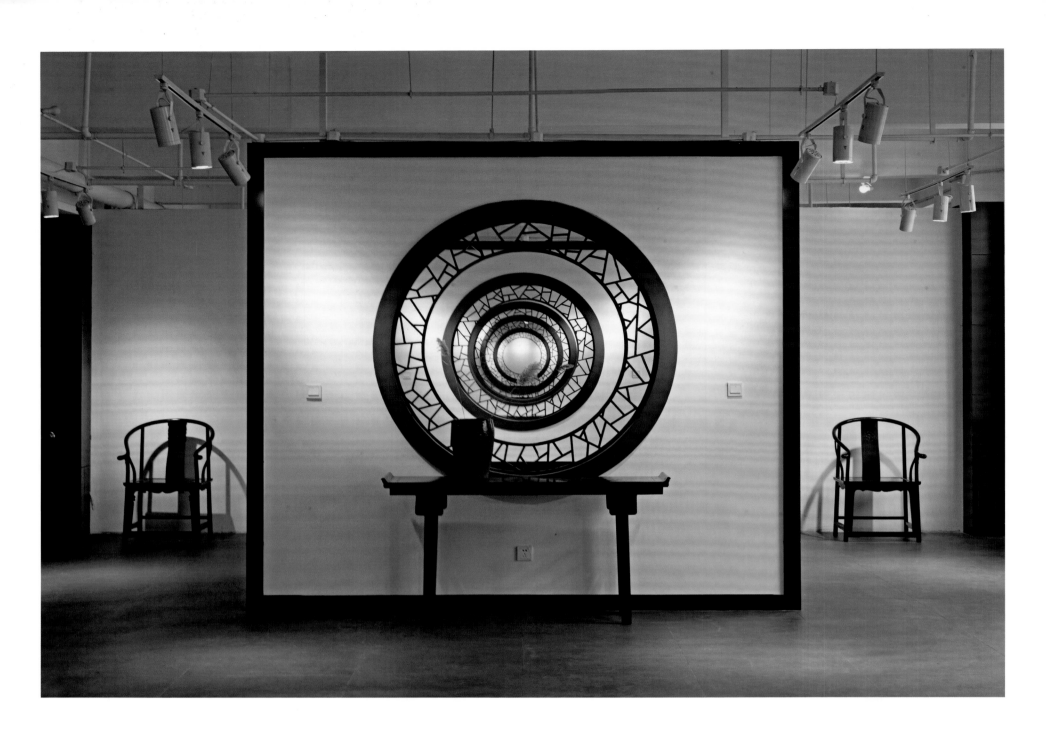

EXPERIENCE PAVILION

21st APIDA

STUDENTS' WORKS
学生作品

STUDENTS' WORKS SILVER MEDAL
学生作品 银奖

Design Team / Chris Lam
Location / Hong Kong, China
Area / 2,000 m²
Client / hospice and palliative care organsation

A FOLDING ARCHITECTURE: THE CENTER OF HOSPICE AND PALLIATIVE CARE

折叠式建筑：临终关怀和姑息治疗服务中心

Designer / Lam Chik Fung - HKU Space

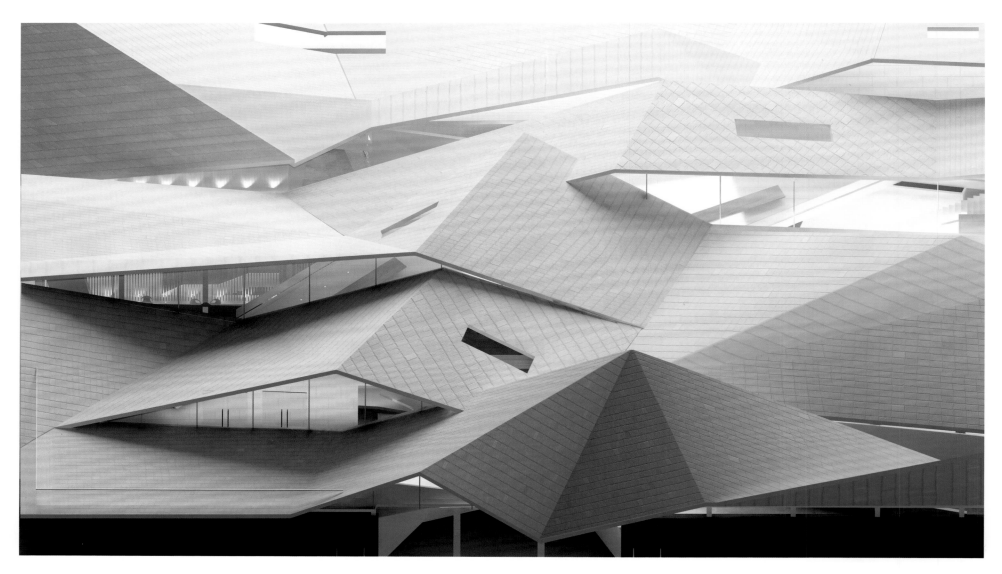

The thesis is to investigate how visitors and the public will handle the issue of death in "randomness". The project aims to promote the importance of afterlife planning, the hospice and palliative care services in Hong Kong.

Inspired by paper crane which means blessings to the death, the project is to design a folding rooftop standing out from the surrounding environment in order to initiate people's emotions. The shape of the exterior does, however, give away the variety of structural distributions within the interior spaces.

All space are defined by folding ceiling which is symbol of "blessing", such as entrance lobby, counselling area, hall and roof plaza. Here, the "counselling" path is wrapped with the folding ceiling.

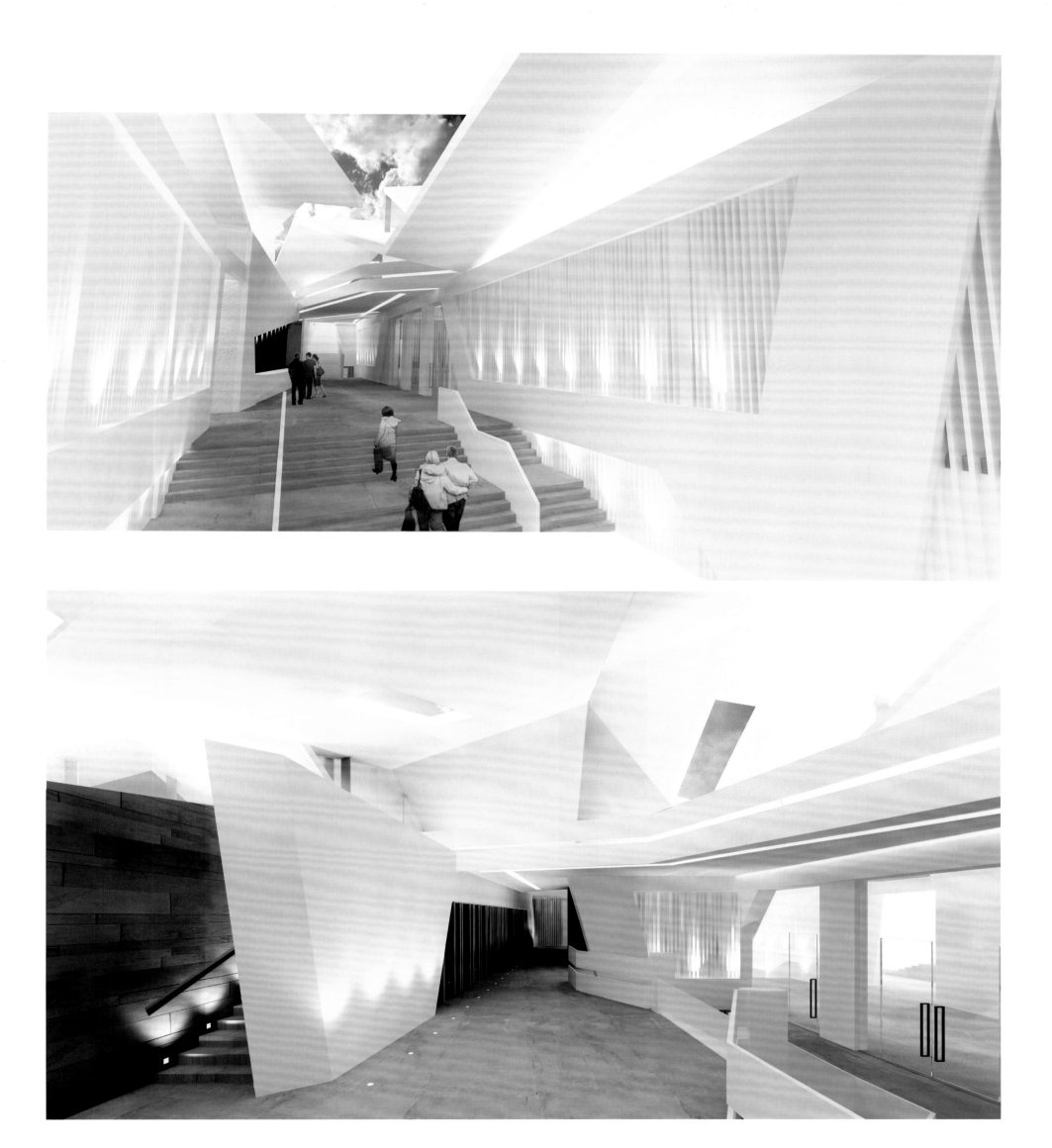

本案是为了调查游客和大众如何对待"不可预测"的死亡。项目旨在在香港地区宣传身后事规划、临终关怀和姑息治疗服务的重要性。

设计灵感来自于纸鹤,承载着对于死亡的祝福,建筑应用了折叠式屋顶,从周围环境中脱颖而出。从建筑的外观可以反映出室内空间结构分配的多样性。

室内所有的空间,例如入口大厅、辅导区、大厅、屋顶广场,全都应用了折叠式天花板,其代表了"祝福"。"安抚"道被折叠的天花板包覆着。

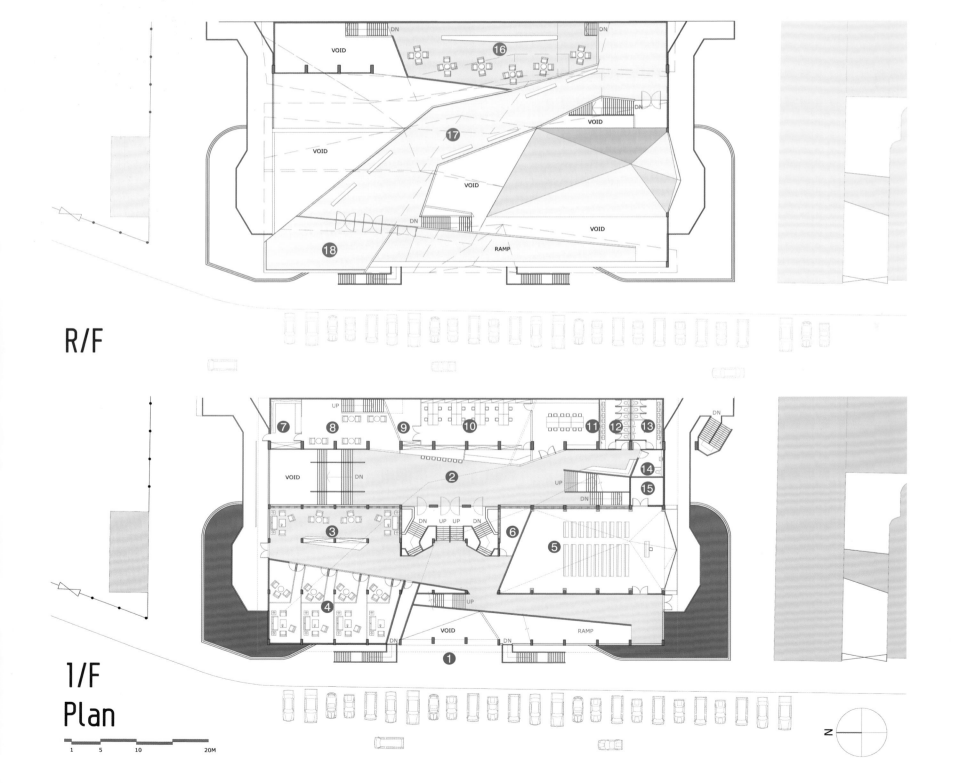

Top View

1 Main Entrance
2 Entrance Lobby
3 Library (counseling area)
4 Counseling Rooms
5 Hall
6 Hall Technical Room
7 Kitchen
8 Cafe
9 Office storage
10 Office
11 Multi-functional Room
12 Female washroom
13 Male washroom
14 Disabled Person washroom
15 Hall storage
16. Cafe (upper level)
17. Covered Roof Plaza
18. Outdoor Roof Plaza

R/F

1/F Plan

A FOLDING ARCHITECTURE:

THE CENTER OF HOSPICE AND PALLIATIVE CARE

STUDENTS' WORKS	BRONZE MEDAL		Location / Hong Kong, China
学生作品	铜奖		Area / 870 m²
			Client / Hong Kong Tourism Board

FLOATING COMMUNITY — H2O HOTEL

漂浮社区——H2O 酒店

Designer / Kane Wong - Birmingham City University

The project proposes to design for a Floating Community Experience Complex raising the public awareness towards rising of sea water level by an active approach to "live with water". In the complex, there is an Exhibition Center for general information, a Water Entertainment Center for a day experience and some Accommodation Suites design for overnight experience of living on water and a Floating Bridge design for the link from land to water.

Site - Aberdeen Typhoon Shelter - As a Historical and Famous Water Community of Hong Kong.

Help to promote and reactivate the floating community culture in HK.

Design Element/Programs

Experience from an hour indoor exhibition center to a day for outdoor water entertainment and finally an over-night in-and-out experience in the accommodation suite.

项目旨在创造出一个漂浮社区体验综合体，通过近似于"与水同居"的方式，来提升公众对于海平面上升的意识。在综合体内，有一个基本资料的展示中心，一日体验的水上娱乐中心，一些用作水上过夜体验的套房，以及一个浮桥，用来将水面和陆地连接起来。

场址——香港仔避风塘——香港历史性的有名的水社区。

促进和再现香港的漂浮社区文化。

设计元素

体验从一小时的室内展览中心到一日的户外水上娱乐，最后是在住宿套房内整晚时进时出的漂浮体验。

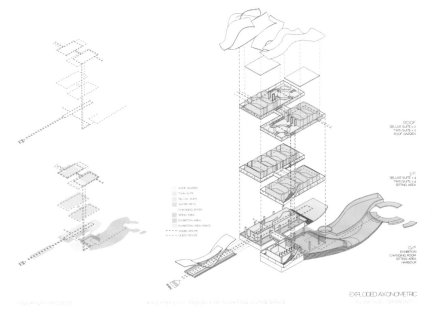

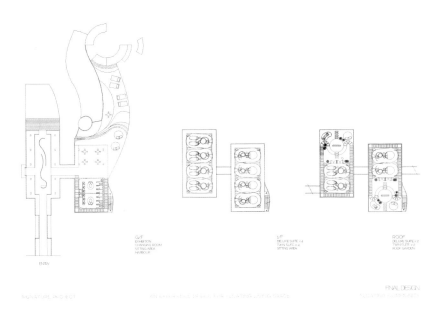

H2O HOTEL

● STUDENTS' WORKS EXCELLENCE MEDAL
学生作品 优秀奖

Location / Hong Kong, China
Area / 5,000 m²

STUDARY

墨坊

Designer / Trisica Nip - Hong Kong Design Institute

Ever since the 6 industries have been proposed by HKSAR, people in Hong Kong have been drawn and waiting for the emerge of cultural and creative industries. However, it is a lack of support for this expertise. The need to call on a full range or latest released references is summoned to equip the design schools in Hong Kong. That leads to the birth of "themed library", which enables campus and non-campus users to access to the essential information without barriers. Studary is a mix of studio and library, a platform for design students to nurture peer learning, operation and fresh ideas.

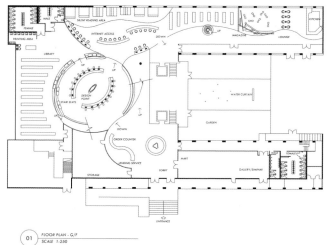

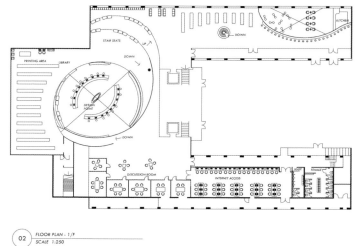

"Zhi shu shi mo" is a Chinese phrase which means to know and acquire knowledge (Mo means ink in Chinese). Library is a place for people learning.

Thus, ink was chosen to be the concept of this library. ●

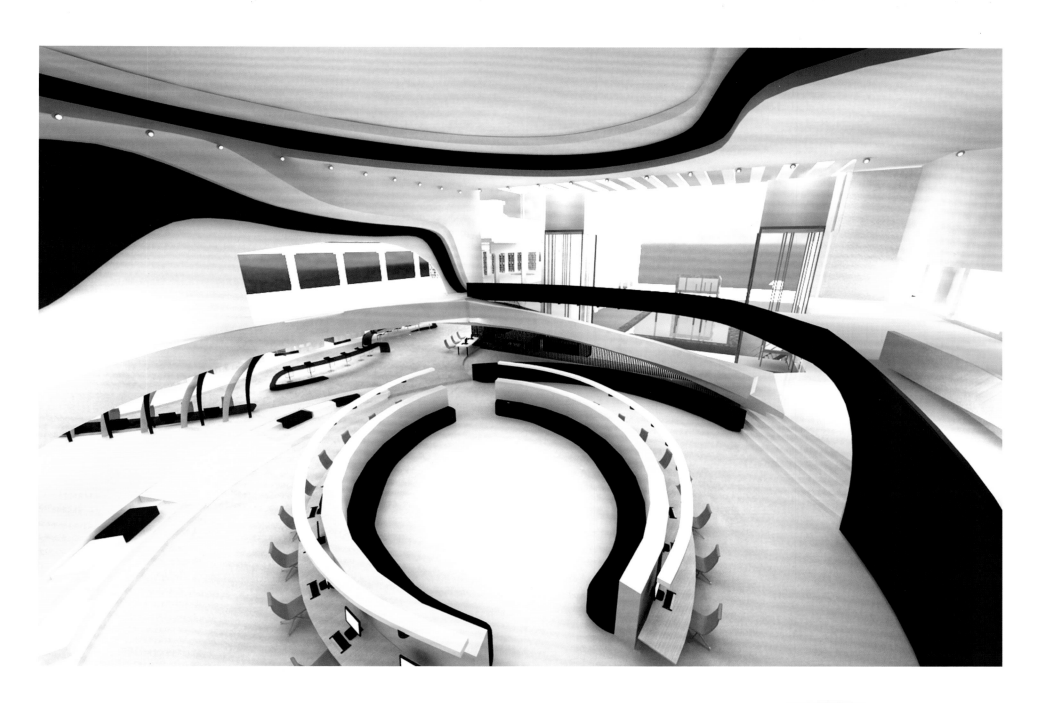

自从香港特区提出了6种工业以来,香港人民就在等待着自己的文化创新产业的出现。然而,香港创新产业缺乏专业知识的支持。因此,为香港地区的设计学院配备提供全方位及最新资讯的机构的需求与日俱增。"主题图书馆"应运而生,可以让学生抑或是非学生用户无障碍地获取重要资讯。"墨坊"是工作室和图书馆的合体,是为设计类学生提供的一个平台,可供他们彼此学习交流,产生新鲜创意。

"知书识墨"是指知道和获取知识。图书馆是人们学习的地方。

因此,墨被选择来作为这个图书馆的设计理念。

STUDENTS' WORKS 学生作品 EXCELLENCE MEDAL 优秀奖

Location / Hong Kong, China
Area / Over 1,500 m²
Client / Prefer Government

MOVABLE BEACH

可移动的海滩

Designer / WONG Tsz Wun Veronica

Hong Kong's image of contaminated modern city belies its wealth of natural green resources. Most of these sites are in close proximity to the city, so visitors can easily reach them within 60 minutes for a pleasurable and educational outdoor experience.

Unfortunately, those green islands are contaminated by urban development day by day.

So, people have to take protective measures!

Movable Beach is a recreation and conservation place to connect people and the islands. Visitors can learn, experience and act in different sites. It will travels to 7 islands in Hong Kong where have no beach available for swimming. Also, there have much wealth of natural green resources. People can have an Eco-tourism and hold the conservation of workshop on the place. Movable Beach can transform into 6 units of triangle shapes. There will provide a glass beach and some exhibitions which used all recycled materials. ●

香港受污染的现代都市形象掩盖了其所拥有的自然绿色资源。这些地方大部分离城市非常近，游客可以在一小时内轻松到达，去享受愉悦的户外体验。

不幸的是，随着城市的发展，这些绿岛也日渐被污染。

因此，我们要采取保护措施！

可移动的海滩是一个休憩保护地，将人们与这些绿岛联系起来。游客可以在不同的地点去学习、了解和体验。其将去往香港的7个岛屿，那些岛屿上没有可供游泳戏水的海滩，却有着自然绿色资源的巨大财富。可以发起生态旅游，召开针对这些地方的环保研讨会。可移动的海滩，可以变形为6个三角形单元。化身为玻璃海滩和展会，设计所用材料全部为回收利用材料。

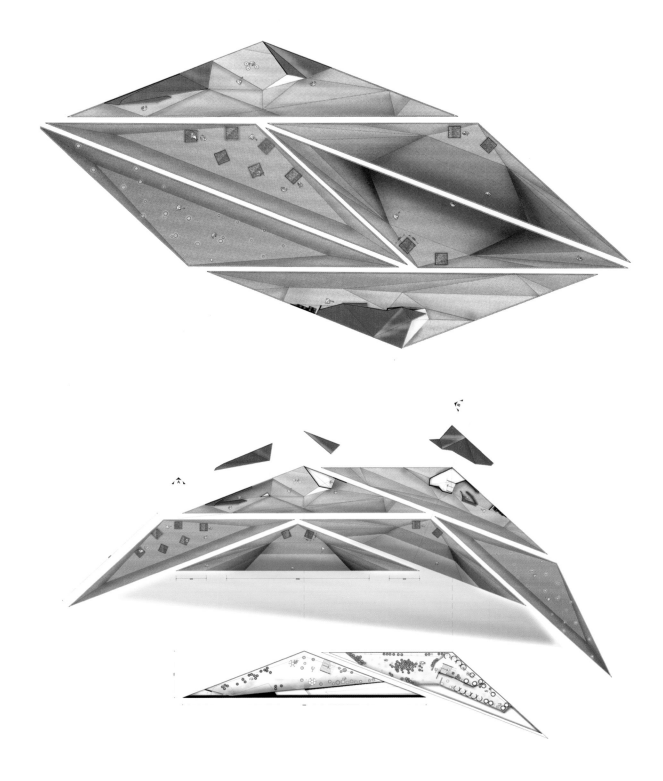

• STUDENTS' WORKS 学生作品   TOP TEN 十名入围

BAND SOUND CENTER

乐队音乐中心

Designer / Wong Wai Hong

Through Café, CD Shop, live stage and musical instruments testing to promote band sound. The center will also provide musical instruments lessons, Band room rent and a large stage. Band team can exchange their opinion in the center. Also band sound center will become the band sound's centrally. Let's band culture popular in Hong Kong.

Music has a relaxing effect, however a dynamic melodies of band sound culture can let people feel a kind of release. So I extract RHYTHM and ECHO these two elements from the band culture to make a space for showing dynamic atmosphere, let people enjoy the band music.

通过咖啡厅、CD店、现场舞台和乐器测试来宣传促进乐队音乐。中心还提供乐器课程、乐队室出租和一个大大的舞台。乐队成员能在此交流意见。乐队音乐中心将成为乐队音乐的集中地,让乐队音乐在香港流行起来。

音乐有着让人放松的效果,乐队音乐有活力的旋律也能让人感到轻松的氛围。设计师从乐队文化中提取了韵律和重复这两个元素,创造出有活力的空间氛围,让人们享受音乐。

Location / Hong Kong, China
Area / 4,500 m²

CENTER

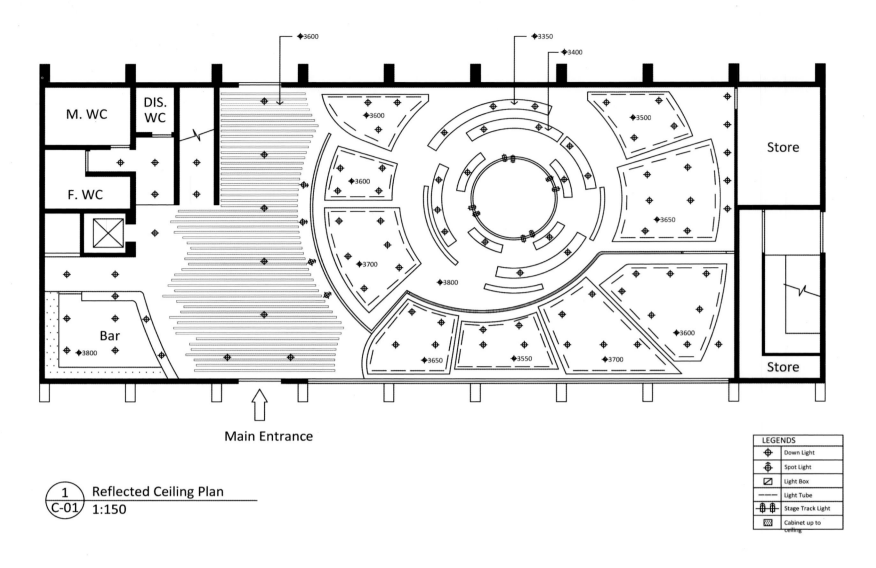

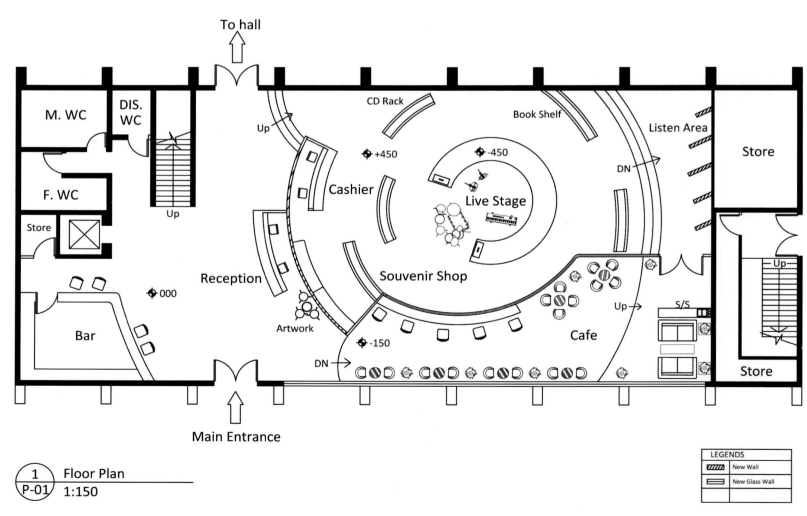

STUDENTS' WORKS | TOP TEN
学生作品 | 十名入围

Location / Hong Kong, China
Area / 465 m²
Client / Hong Kong Baptist Church

CHURCH

教堂

Designer / Wong Mei Mei

Church is a holy place. The Christian share the word of God and worship together in the chapel. Church is also a place that brings positive energy and the other home for the Christian. Therefore, Church should be warm, cheerful and witness our life story.

The design concepts are Holy, Peaceful, and Lively. It is inspired by water. First of all, designer uses the natural elements such as sunlight, water and plants to present the mood in holy and peaceful. It is because the sunlight can make people feel warm and give them positive energy. When the sunlight enters the hall, the narrow and high windows will form the shadow on the bench. When people are there, it will seem that the twilight is shining on their face and body. Besides, water is very important in the Baptist church since water means to wash away the sins. Also, water can make people feel comfortable and relaxed. Designer uses different state of water to bring the peaceful, comfortable and the relaxed mood. For examples, water waves, waterfalls and the shadow of water. What's more, the plants will be used to present the lively and the thankful for the God. Lastly, designer uses many colorful elements to bring happy feeling and be attractive.

　　教堂是神圣的地方。基督徒在此共聚，尊崇和分享圣言。教堂也是一个传达正能量的地方，是基督徒的另一个家。因此，教堂应该是温暖的，愉悦的，是生活的见证。

　　本案的设计理念是神圣、和平和活力。设计灵感来自于水。首先，设计师使用了自然的元素诸如太阳光、水和植物，来代表神圣平和的情绪。阳光能让人感觉温暖，给人正能量。阳光照入大厅，细高的窗户在长椅上投下影子。当人坐下时，这些影子则会出现在他们的脸上和身上。另外，水在浸信会教堂中非常重要，因为水可以洗去罪恶。同时，水能让人感觉舒适和放松。设计师想要使用不同形态的水，例如，水波纹、瀑布和水的影子，来表达和平、舒适和放松的氛围。此外，植物的使用代表着生机勃勃，以及对于主的感恩。最后，设计师使用了多彩的元素来创造愉悦的气氛，增强空间的吸引力。

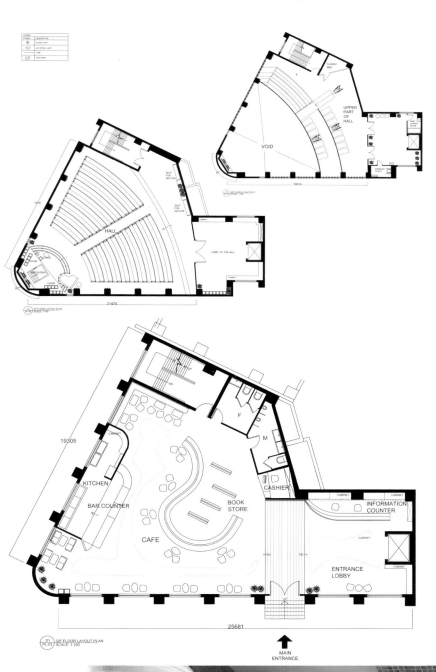

YOUTH HOSTEL

青年旅社

Designer / Edwin Tan

This hostel for youth used a theme "origami" as its main design concept. When you think of the traditional Japanese art form origami, you might imagine visions of paper cranes, flowers and elaborate creations that are often small enough to fit in the palm of your hand. The origami style is being transformed into new, fun, decorative and exciting forms that can add a touch of modern playfulness to your interior design. And it can represent fun and playfulness of youth, because youth are usually playful, like to bring in the fun, and have a way of lightening up any social situation, with jokes or funny stories sometimes. This is the reason that designer chooses the concept.

Location / Malaysia
Area / 914.1869 m²
Client / Limkokwing University

本案是一间青年旅社，主要的设计理念来自于"折纸"。当你想到传统日本艺术形式折纸的时候，你可能会联想到纸鹤、纸花以及其他一些小小的能放在手心的精致折纸工艺品。折纸被转变成一种全新的、有趣的、装饰性的、令人兴奋的形式，给室内设计带来一种现代而具有趣味的格调。其能代表年轻人的趣味和玩乐，因为年轻人通常都爱玩，有时喜欢用玩笑和故事的形式营造轻松的氛围。这也是选择折纸作为这间青年旅社设计理念的原因。

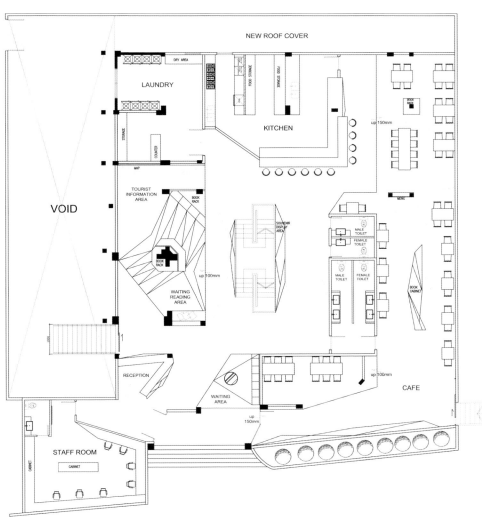
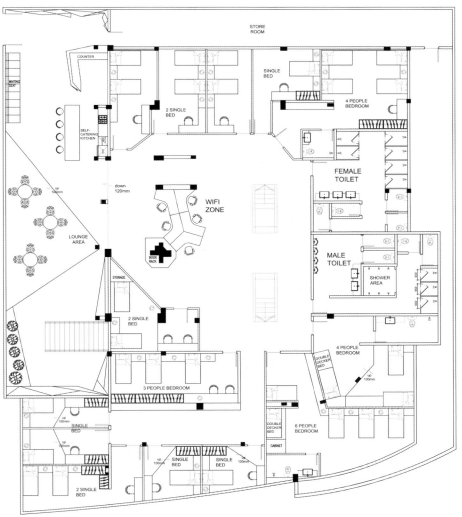

HOSTEL

| STUDENTS' WORKS | TOP TEN |
| 学生作品 | 十名入围 |

PLEASURE LEARNING

寓学于乐

Designer / Wong Wai Hong

Use relaxed, enjoy and pleasure to learn the knowledge to create the theme. Let them to enjoy the atmosphere of learning, to keep learning and further education or job training, and absorb all kinds of knowledge. Let them to get rid of the boring learning atmosphere and concept, to attract the people who stay in this space is easier to enjoy the learning process and take the initiative to draw a variety of knowledge or play interest. ●

　　本案的设计主题是以轻松愉悦的方式来学习知识，让人们享受学习的氛围，不断学习深造，吸收各种知识。让人摆脱枯燥的学习氛围和概念，吸引身处这一空间的人们，使其享受学习的过程，主动去吸取各种知识。

Location / Hong Kong, China
Area / 10,000 m²
Client / Sun Hung Kai & Co Ltd

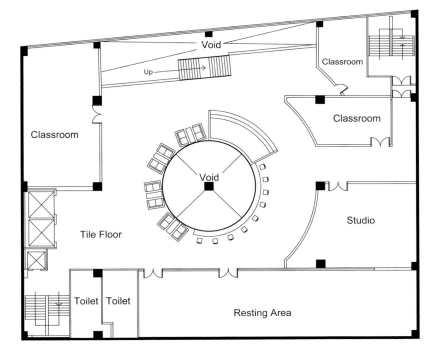

1/F Floor Plan
1:200

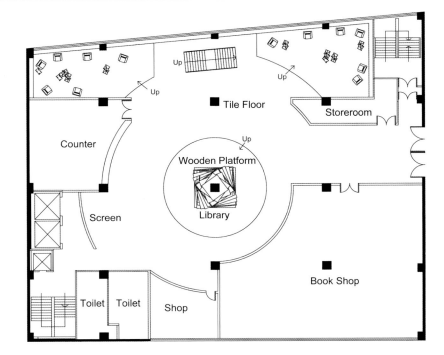

G/F Floor Plan
1:200

- STUDENTS' WORKS 学生作品
- TOP TEN 十名入围

Design Team / Kasey Lam
Location / Hong Kong, China
Area / 1,940 m²
Client / Simply Mind

REFRESHING EXPERIENCE IN MIND & BODY HEALTH CLUB

身心健康的全新体验俱乐部

Designer / Lam Wing Sze

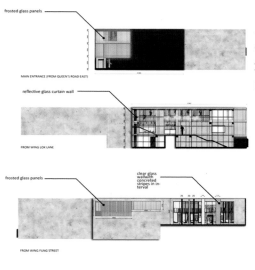

"ENTRANCE AT THE CORNER FOR BETTER ACCESS TO ANOTHER VOLUME OF THE BUILDING." - the exterior

"FROSTED GLASS WALL PANEL SYSTEM BECOMES A GIANT LIGHTBOX OVERHEAD, WITH A SERIES OF DECORATIVE HANGING LAMP ON CEILING." - the lobby

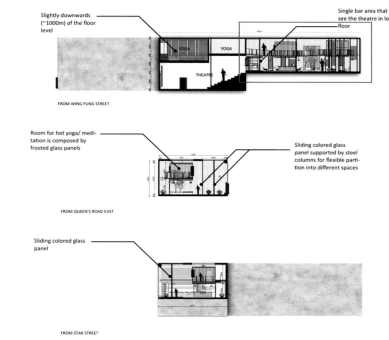

"TWO FLOATING ROOMS MADE BY FROSTED GLASS WALL PANEL FOR HOT YOGA & MEDITATION. THE ROOMS IS LIGHTENED UP BY SUNLIGHT AT DAYTIME." - the coffee zone

Negative image of medical or clinical like mental health center always distracts people to visit for help. To make the visitors feel comfortable or even feel proud to visit, interior design participates by refreshing their perception on mood, considering mental health as regular body fitness and professional mind training activity.

By refreshing the experience and altering the nature of mental health center into lifestyle mind & body health club, the club is for the professionals who strive for spiritual satisfaction. Simply Mind is a brand new health club where you can experience healthy spiritual, physical and social lifestyle through mind training & soul strengthening.

Inspired by lightweight, visually lightweight is able to provide a relaxing & welcoming atmosphere for further enjoyment of mind, body & social activities inside the club. This concept could be expressed through four design elements: Material; Natural lighting; Pattern; Color. ●

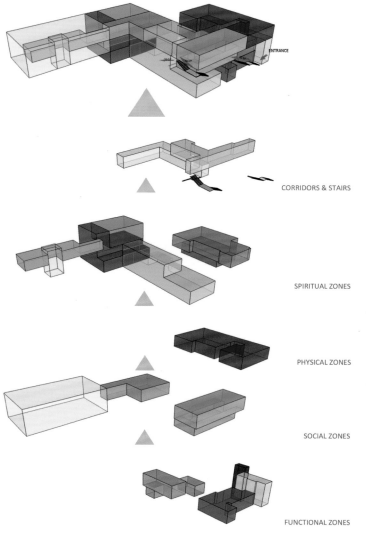

ENTRANCE

CORRIDORS & STAIRS

SPIRITUAL ZONES

PHYSICAL ZONES

SOCIAL ZONES

FUNCTIONAL ZONES

人们通常不愿去给人负面意象的医疗诊所寻求帮助，例如心理健康中心。为了让访客感觉舒适，愿意来此，本案利用室内设计来改变这些场所给人的感知印象，让保证精神的健康成为和保持身材以及进行职业能力培训一样的行为。

将心理健康中心转变成身心健康保健俱乐部的性质，是为寻求精神健康的人士所设计。人们可以在此进行思维的训练和精神强化，体验健康的精神状态和社会生活方式。

设计灵感来自于轻盈，视觉上的轻盈感能带来放松的舒适氛围，让人在会所中进一步得到思想、身体和社交活动的愉悦享受。这种理念可以通过4种设计元素来表达：材料、自然光、图案和色彩。

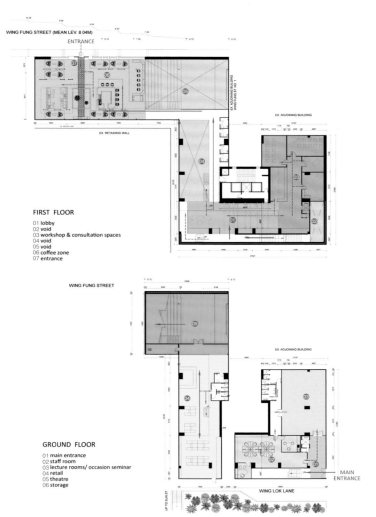

FIRST FLOOR
01 lobby
02 void
03 workshop & consultation spaces
04 void
05 void
06 coffee zone
07 entrance

GROUND FLOOR
01 main entrance
02 staff room
03 lecture rooms/ occasion seminar
04 retail
05 theatre
06 storage

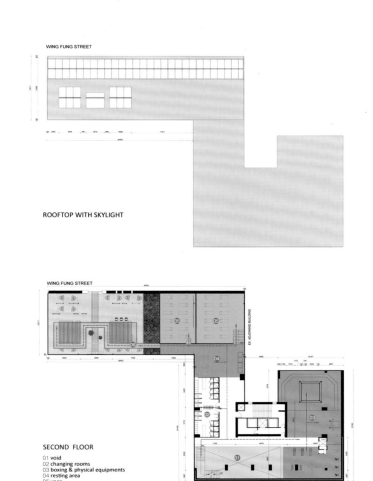

ROOFTOP WITH SKYLIGHT

SECOND FLOOR
01 void
02 changing rooms
03 boxing & physical equipments
04 resting area
05 yoga
06 hot yoga & meditation

BODY HEALTH CLUB

21ˢᵗ Asia-Pacific Interior Design Awards 2013

Into the 21st year, APIDA continues to give recognition to our outstanding interior design projects and designers, promoting professional standards and ethics among interior design practices operating in the Asia Pacific region. The APIDA Program this year attracted almost 600 entries from the Asia Pacific Region which were judged by both the local and international panel of prominent designers, architects and academics to select the Winners and Excellences in the ten categories.

APIDA Aims:

- to promote public awareness of interior design as an important aspect of everyday life;
- to acknowledge / give industry recognition to deserving projects and designers;
- to encourage and promote professional standards and ecthnics among interior design practices operating in the region

APIDA's Judging Criteria:

- Originality and Innovation
- Functionality
- Space Planning
- Aesthetics

Awards categories:

ONE Gold, ONE Silver, ONE Bronze and TWO excellences for each category.

ASIA PACIFIC INTERIOR DESIGN AWARDS 2013
亞太區室內設計大獎　二零一三年

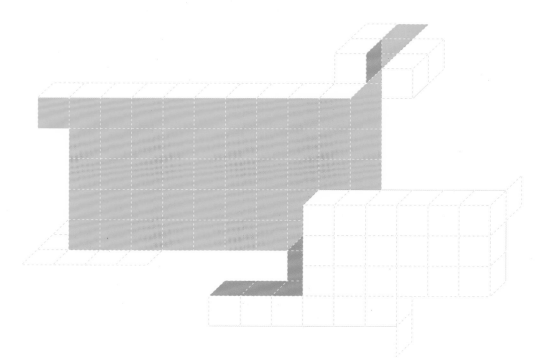

www.acs.cn

ACS
Artpower Creative Space 创意 | 空间

国际视野新站点 | 案例丰富新颖 | 访谈顶尖设计 | 挖掘新锐设计师 | 国际设计界缩影

ACS 创意·空间　　ABOUT

十年专注于建筑、室内、景观和平面设计，业务横跨图书出版、发行、文化传媒、品牌运营及艺术品市场等多个经营领域，Artpower 自版发行 600 多本图书，收揽全球顶尖设计公司和设计师近 40000 套优秀原创作品（不断更新ING）。

ACS 整合 Artpower 线上资源，推荐前沿创意理念、概念性设计思维；发布创意赛事活动；组织设计大师访谈；展示新锐设计师作品，推介设计项目；提供私人定制出版和众筹出版等服务。与国际设计团队对接，在全球范围内打造专业设计师展示和交流平台。

我们能做什么？　　HOW

注册成为网站会员，做主个人网页，独享会员特权；上传个人作品，展示设计理念，交流成长，互通合作契机。

登录浏览，尽享 40000+ 海内外设计大师作品；建筑、室内、景观、平面、产品、环境设计等分门别类，轻松导航，应有尽有。

挖掘新锐设计师　　DESIGNERS

· ACS 线上展厅
我的 ACS 我做主！设计师可以尽情发表自己的作品，让世界各地的设计师共同关注你的成长！

· 设计师发布会
如果你还在为身为"新人"的标签所困扰，ACS 展示平台只有"新锐设计师"。把曾经因各种原因被否掉的方案重新发表出来，也许你就是那个我们要找的设计师！

· ACS 把设计师的项目推送给全世界，设计无国界，一起关注和交流！

DESIGN FOR DESIGN

IMAGINATION — CREATIVITY — ABILITY

私人定制出版　　PRIVATE

ACS 创意空间联营平台为您提供私人定制出版服务。

线下俱乐部　　ACTIVE

ACS 创意空间俱乐部，不定期邀请国内外顶尖设计师，举办各种创意设计讲座、创意沙龙等，分享天马行空的有趣创意，是思维碰撞、灵感横溢的场所，是趣味相投、惺惺相惜的交友平台，也是企业品牌的展示空间。有机会成为线下俱乐部盟主！

设计 · 杂志 · 中英文
Artpower Creative Space（ACS）创意空间（245mm×325mm · 168 页 · 68CNY）

《ACS 创意空间》杂志是 Artpower 倾力打造的高端空间设计专业期刊，中英双语，全球同步发行；单期发行量逾万册，更有黎巴嫩等国家的专售版；装饰行业至佳交流平台，传播设计新锐资讯；高端空间设计专业期刊，发布国际最优秀室内设计师和建筑设计师的最新作品。

深圳市艺力文化发展有限公司
艺力国际出版有限公司（香港）
深圳市艺力文化发展有限公司北京分公司
深圳市艺力文化发展有限公司厦门分公司

出版合作 / 广告合作：rainly@artpower.com.cn（王小姐）
作品投稿：artpower@artpower.com.cn（莫小姐）

艺力 ACS 创意空间
扫描即可关注！

深圳市米兰映象工艺品有限公司

成立于2005年,原名创美家饰。产品曾出口欧美十几个国家和地区。米兰映象创始人阳小美先生室内设计专业毕业,开了近十年室内设计公司,极其爱好家居产品设计。公司注重产品研发、品质与服务,强大的研发团队与严格的品质管理流程,受到业内广大设计师好评。公司专业为五星级酒店、高端会所、房地产样板房及其他商业空间提供多种风格艺术挂画、摆件、雕塑等设计定制整体配饰服务。

MILAN IMPRESSION
米蘭·映象

引领家居时尚,传递生活美学
LEADING HOME FURNISHING FASHION, DELIVERING THE LIFE AESTHETICS

▶ 自主生产

公司旗下自有陶瓷、金属、木雕、制画装裱等工艺三大工厂,七大品牌,一万多种不同风格极具设计感的产品品类,厂房面积超过20000平米,拥有员工200余人。在全国最大的家居饰品集散地艺展中心,公司拥有7个展厅,超过1000平米展示面积。

关注米兰映象·发现生活美学

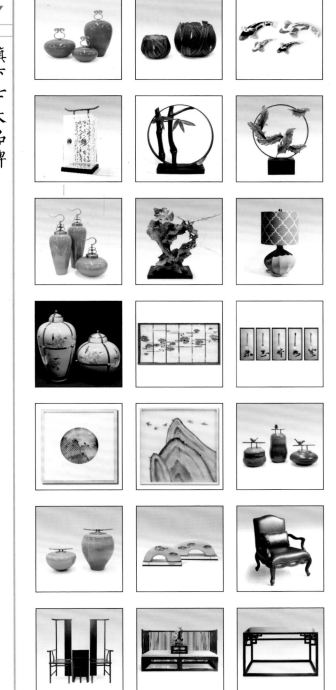

旗下七大品牌

| 定制服务 |

我们提供定制服务，公司拥有各大美院雕塑专业人员多名，如果您需要定制产品，请您提供以下内容：图片、尺寸、颜色、材质、工艺要求、边框（装饰画），我们会竭尽所能实现您的设计梦想。

服务热线：**400-1000-776**　客服QQ：**2100588790**

工厂地址：深圳市龙岗区同乐三棵松工业园

深圳市米兰映象工艺品有限公司
www.milanimpression.com.cn